How to Restore & Improve

Classic Car Suspension, Steering & Wheels

Repair • Restoration • Maintenance

Enthusiast's Restoration Manual Series

Beginner's Guide to Classic Motorcycle Restoration, The (Burns)
Citroën 2CV, How to Restore (Porter)
Classic Large Frame Vespa Scooters, How to Restore (Paxton)
Classic Car Bodywork, How to Restore (Thaddeus)
Classic British Car Electrical Systems (Astley)
Classic Car Electrics (Thaddeus)
Classic Cars, How to Paint (Thaddeus)
Classic Car Interiors, How to Restore (Steinfurth, Zoporowski)
Classic Off-road Motorcycles, How to Restore (Burns)
Honda CX500 & CX650, How to Restore – YOUR step-by-step colour illustrated guide to complete restoration (Burns)
Honda Fours, How to Restore – YOUR step-by-step colour illustrated guide to complete restoration (Burns)
Reliant Regal, How to Restore (Payne)
Triumph TR2, 3, 3A, 4 & 4A, How to Restore (Williams)
Triumph TR5/250 & 6, How to Restore (Williams)
Triumph TR7/8, How to Restore (Williams)
Triumph Trident T150/T160 & BSA Rocket III, How to Restore (Rooke)
Ultimate Mini Restoration Manual, The (Ayre & Webber)
Volkswagen Beetle, How to Restore (Tyler)
VW Bay Window Bus, How to Restore (Paxton)
Yamaha FS1-E, How to Restore (Watts)

General

Anatomy of the Classic Mini (Huthert & Ely)
Which Oil? – Choosing the right oils & greases for your antique, vintage, veteran, classic or collector car (Michell)

www.veloce.co.uk

First published in April 2018 by Veloce Publishing Limited, Veloce House, Parkway Farm Business Park, Middle Farm Way, Poundbury, Dorchester DT1 3AR, England. Tel: 01305 260068 / Fax 01305 250479 / e-mail info@veloce.co.uk / web www.veloce.co.uk or www.velocebooks.com.

ISBN: 978-1-787111-87-5 UPC: 6-36847-01187-1.

Originally published as *Praxishandbuch Fahrwerk & Räder, Theorie – Technik – Optimierung* by HEEL Verlag GmbH, Königswinter, Germany. Edited by HEEL Verlag: Tobias Zoporowski, Jürgen Schlegelmilch. Produced in cooperation with and courtesy of *Oldtimer Markt* magazine & Peter Steinfurth. Cover photo: Jürgen Schlegelmilch. Photo credits: *Oldtimer Markt* Archive. English translation: Julian Parish.

Readers with ideas for automotive books, or books on other transport or related hobby subjects, are invited to write to the editorial director of Veloce Publishing at the above address. British Library Cataloguing in Publication Data – A catalogue record for this book is available from the British Library.

Typesetting, design and page make-up all by Veloce Publishing Ltd on Apple Mac. Printed in India by Parksons Graphics.

How to Restore & Improve

Classic Car Suspension, Steering & Wheels

Repair • Restoration • Maintenance

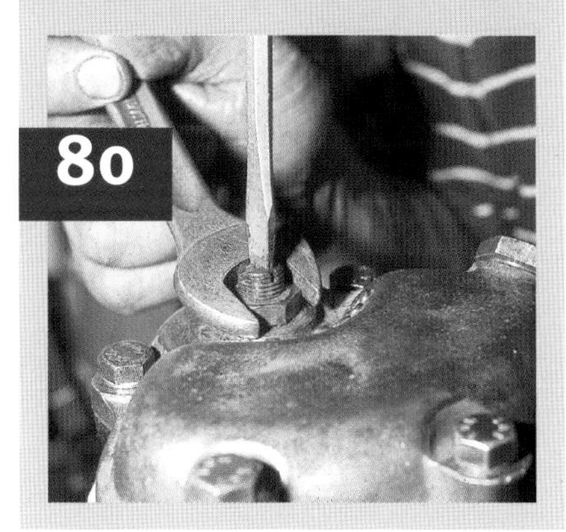

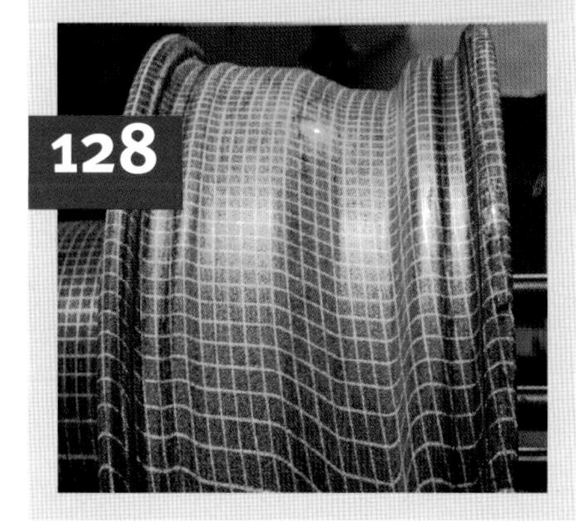

A car's running gear is what connects its body to the road. The condition of all the chassis and suspension components makes the difference between running off the road, with the car out of control, and arriving safely at your destination. You should, therefore, pay particular attention to these parts, even though they are often hidden from view. This is all the more important when your car is old enough to become a classic.

You should be aware that the running gear of a car is not just a matter of springs and shock absorbers. This term also covers everything from the connecting links to the axle to the contact patch of the tyres. These should all be checked regularly and, if necessary, replaced to ensure that the car stays safely on the road.

All of these components are discussed in this manual, and the coverage will lead you to an understanding of the role they play. The book also looks at the different types of steering mechanism, as well as power-assisted steering, although these topics almost fall outside its scope. You will learn about refurbishing ageing wheels, and why you should not do everything which is technically feasible. In addition, you will pick up some valuable information about tyres for classic cars. The book will show you where each part goes, how to repair damage yourself and what services are offered by specialists. Note that the manual deliberately omits any coverage of brakes; this could easily fill another restoration manual on its own.

Whether you'd like to set up your car for sportier handling, or simply want to refurbish its worn-out suspension parts, this restoration manual should serve you well.

Disclaimer

The specialist advice in this book assumes that the reader will have sufficient technical knowledge and mechanical ability to use it safely; if that is not the case, you should not take on these jobs. If in doubt, always consult a trained specialist for advice before trying to do it yourself. If used ineptly, some of the technical advice given may, in certain circumstances, result in serious damage or personal injury.

All advice contained in the book is given in good faith, but the authors, translator and publisher expressly decline all responsibility for the instructions published here and the results which may arise from following them. The reader takes full and sole responsibility for the risks which he or she incurs. The responsibility of the authors, translator and publisher is also expressly excluded for any printing or spelling mistakes or factual errors of any kind.

Currency guide

Throughout this book prices of parts, components and procedures are given by the ● symbol. At the time of writing the currency symbol equals €1.00/£0.89/$1.23. Please adjust to suit current exchange rates.

Foreword

160

Theory
Chassis and suspension: the basics

The main purpose of a car's suspension is to keep all four wheels on the road: nothing more, nothing less. Even the best tyres can provide precious little roadholding if they are left hanging in mid-air.

Over the past hundred years, all manner of approaches have been taken to solve this apparently simple problem, with varying degrees of success.

What a shame that there isn't a version of 'Top Trumps' for car suspension components. I can just imagine how a Ferrari 250 GTO might come up against a Citroën 2CV, and the child with the card for the 2CV would proudly read out: "Independent rear suspension, horizontally mounted coil springs and telescopic shock absorbers – aha!" Meanwhile, the kid with the card for the Ferrari would look down sadly and say: "Live axle with leaf springs, dual trailing arms and a Watt's linkage – blast!" One reason why this version of 'Top Trumps' will never exist is that the ability to judge the performance of a car's suspension is at risk of dying out.

Is a live axle with leaf springs really a bad thing, and is independent suspension automatically better? In this chapter, we would like to get to the bottom of this question. To understand this more deeply, you need to be familiar with two important concepts, which will already give you an edge over most bar-room pundits. The first of these is the car's centre of gravity, the second is the roll centre.

While nearly everyone can understand the notion of the centre of gravity, the concept of the roll centre is considerably harder to grasp. Roughly speaking, it is the point about which a car leans when cornering, and its location is directly dependent on the suspension design. Since a car has two axles, however, there are, as a result, two roll centres: one on the front axle and one on the rear. If both points are joined by an imaginary line, the so-called 'roll axis' can be obtained.

The higher the car's centre of gravity in relation to this longitudinal axis, the greater the leverage the car's weight will apply to the roll axis and the more the car's body will lean to one side. Now you would think that the suspension designer could simply locate the roll centre for each axle higher up, to reduce the extent of roll. Ultimately, if the roll axis ran through the car's centre of gravity, the degree of roll should be equal to zero. But this theory has one major snag: the car's body would no longer roll to one side, the car would simply overturn. Or, to put it another way: the higher the roll axis, the less weight is applied to the inside wheels when cornering. Even if the car doesn't actually overturn, since the outer wheels would slide first, locating the roll axis too high is undesirable, as the weight of the car is unevenly distributed. Ideally, the weight should be consistently distributed across all four wheels.

In practice, this ideal is impossible to attain, as the car's centre of gravity generally does not lie exactly between the wheels. Cars with front-wheel drive or a rear-mounted engine above all are – in the literal sense of the term – unbalanced in this respect. Since the differing loads on the front and rear axle are usually unavoidable for good reason, the chassis engineers therefore concentrate on maintaining, as far as possible, a constant load on the wheels of at least one axle. When cornering, this comes down to keeping as much weight as possible on the inside wheels. And that in turn means that the roll centre of each axle should be as low as possible.

You are probably wondering by now where this sinister-sounding roll centre is to be found on your classic.

If your car has solid axles at front and rear, it's a simple matter: the roll centre will be at exactly the height at which the springs – in this case, probably leaf springs – work on that axle. This is, moreover, the reason why leaf springs fitted to cars generally run

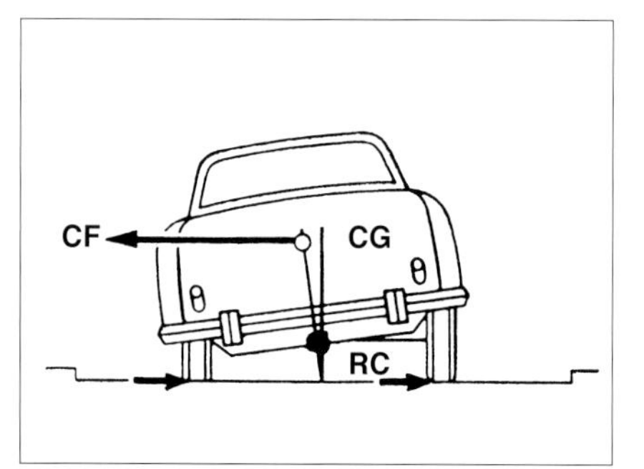

The centrifugal force (CF) allows the car's centre of gravity (CG) to act as a lever on the roll centre (RC).

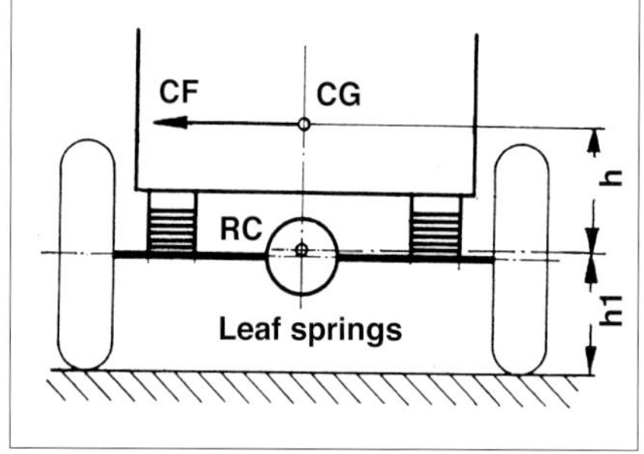

With a live rear axle with leaf springs, the roll centre is situated at the height at which the springs work on that axle.

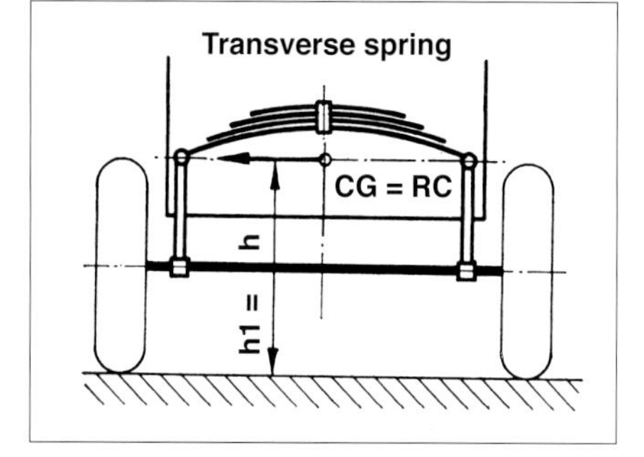

The springs should preferably be located above the roll centre. The downside to this is that the load on the inner wheels will be reduced.

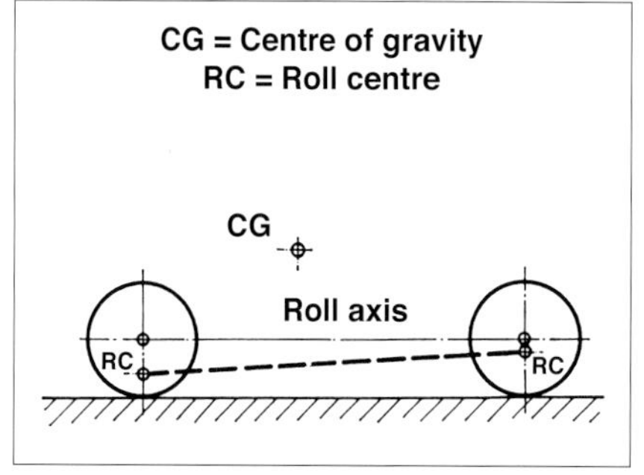

The roll centres of each axle together form the roll axis, about which the car's body leans when cornering.

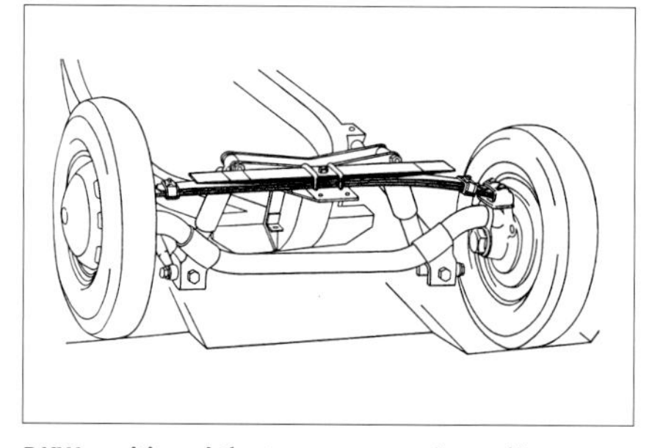
DKW positioned the transverse springs of its 'floating axle' very high up. This reduced roll in corners, but unfortunately not the risk of the car overturning.

A 2CV or a Ferrari – which has the better chassis? The answer depends on the road surface. The Italian car's leaf-sprung live axle has considerable disadvantages due to its weight, while the 2CV corners 'on its door handles.'

underneath the axle. In this way, the roll centre will fall below the mid-point of the axle.

With swing axles, things are not so straightforward. In this case, the number of joints and their layout play a decisive role, but other factors – such as the size of the wheels – also play a part. With trailing-arm suspension, as with the VW Beetle or the 2CV mentioned in the introduction, the roll centre lies at the level of the road surface, which, on the one hand, generates considerable body roll, but also ensures an ideal load distribution on each wheel. Does the 2CV therefore have the ideal suspension? No, since the simple suspension of the little French car completely overlooks one other important factor: namely, the position or alignment of the wheels in relation to the road surface. The 2CV's wheels lean along with its body, until the car is running on the very edge of its wheel rims.

The live axle of the Ferrari 250 GTO does a much better job. Both rear wheels sit firmly upright on the tarmac, regardless of how much the lightweight aluminium body wants to lean. So is the live axle better than its reputation? As long as the GTO is driven as its makers intended, on the smooth surface of a racetrack, then definitely yes. But on a bumpy road like a rally stage, this heavy but otherwise sturdy design will become unsettled. The rear differential, which, of course, makes no contribution whatsoever to the car's roadholding, means that the rear axle – due to its inertia – is really shaken up when driving on rough surfaces, causing it to bounce and suffer axle tramp. Which brings us to a fact known to all chassis engineers: only a wheel that is in contact with the road surface can transmit the forces of acceleration, braking and cornering. This is a subject which we will develop further in the chapters devoted to springs and shock absorbers.

If excess inertia is bad for a car's roadholding, the

A young David Piper proves here that you can take the corners at a decent pace with the rock-hard live-axled chassis of a 1933 MG J4. It is interesting to compare this model with the 1934 MG R-Type (see next page), which was fitted with front and rear independent suspension.

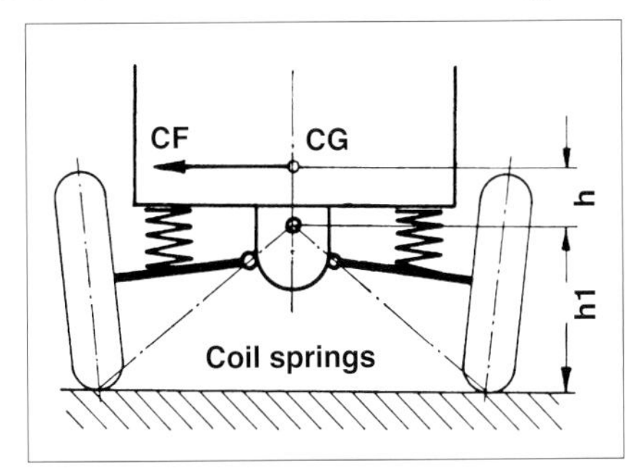

With double-jointed swing axles, the roll centre is at the intersection of the lines running through the mid-point of the tyre contact patch and the axle joints. Much better: the single-jointed swing axle with its low-set pivot point ('Fintail' Mercedes).

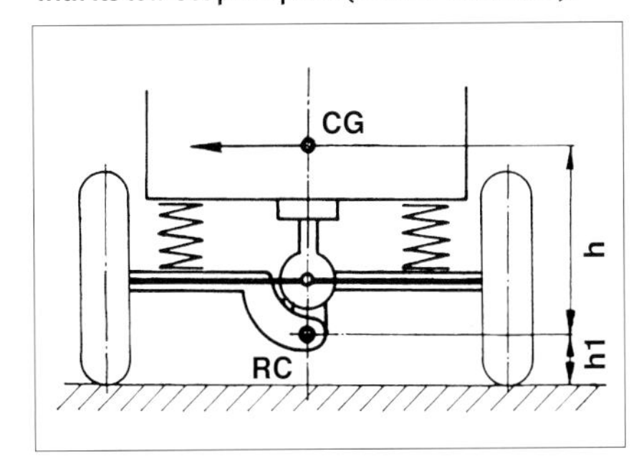

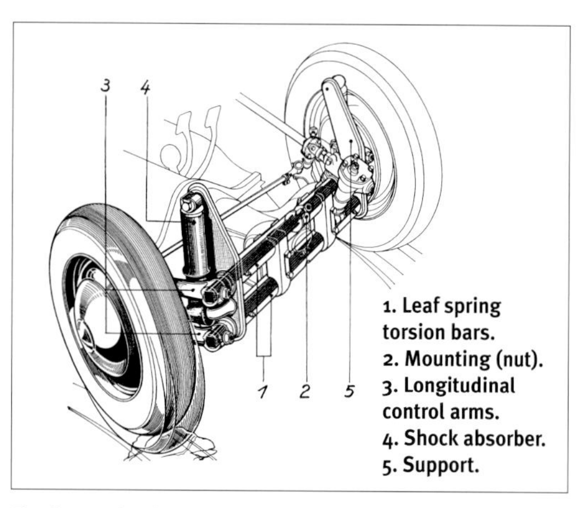

The front axle of a VW Beetle: two longitudinal control arms ensure that the roll centre is kept low, the camber remains constant, and the suspension is light in weight.

Lancia Aurelia: this is what a rear axle might have looked like in 1950, if money were no object. The rear-mounted transmission, brakes and clutch were bolted to the body in order to reduce the unsprung weight, while the movement of the wheels was controlled by the world's first suspension with semi-trailing arms.

1934, and double wishbones already! But MG's engineers had yet to control the R-Type's roll.

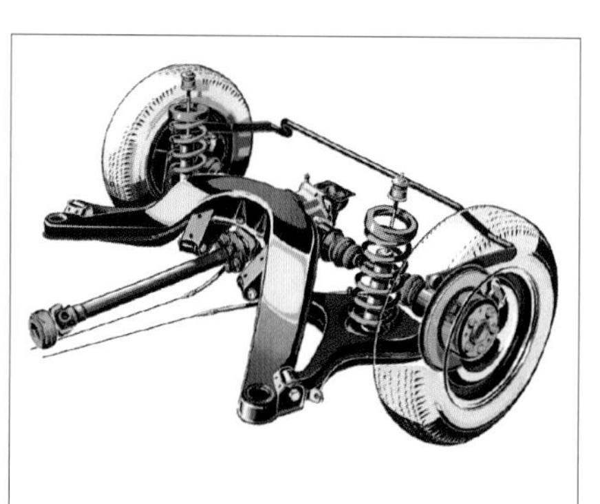

BMW also placed its bet on the semi-trailing arm design, which produced negative camber under load.

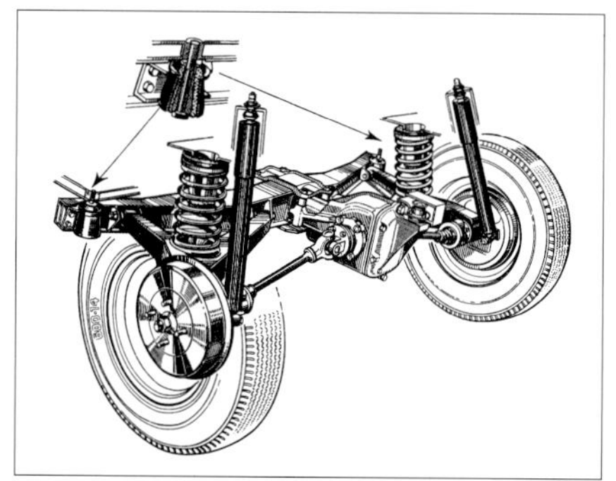

The semi-trailing arm axle with a separate housing for the differential soon gained acceptance.
The exposed propshaft made this design more temperamental than the simple live axle. Illustration: Mercedes 'Stroke Eight.'

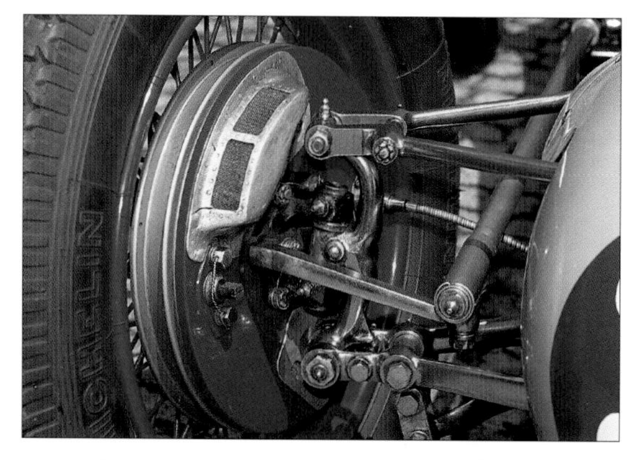

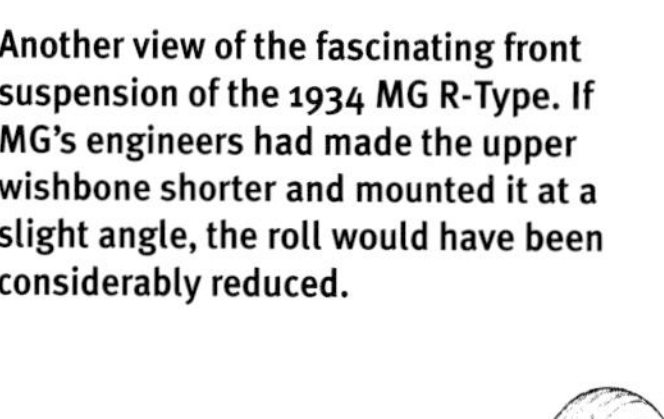

Another view of the fascinating front suspension of the 1934 MG R-Type. If MG's engineers had made the upper wishbone shorter and mounted it at a slight angle, the roll would have been considerably reduced.

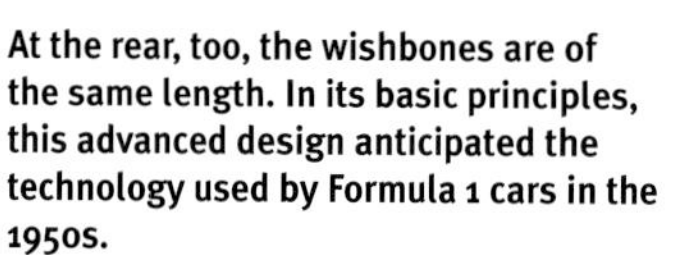

At the rear, too, the wishbones are of the same length. In its basic principles, this advanced design anticipated the technology used by Formula 1 cars in the 1950s.

In a double-wishbone suspension, the position of the roll centre is dependent on the roll angle. The lower wishbone should be horizontal (right).

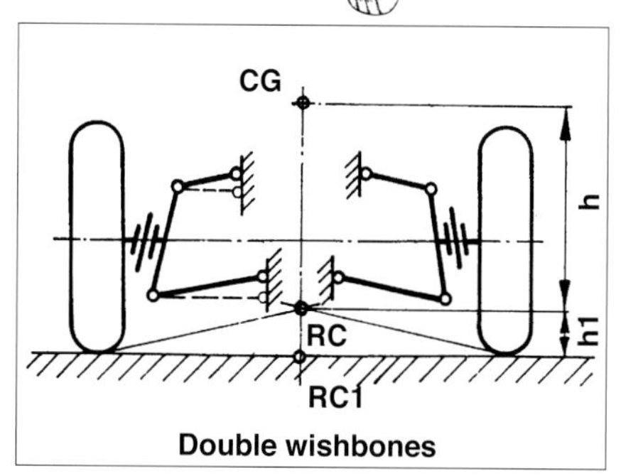

Double wishbones

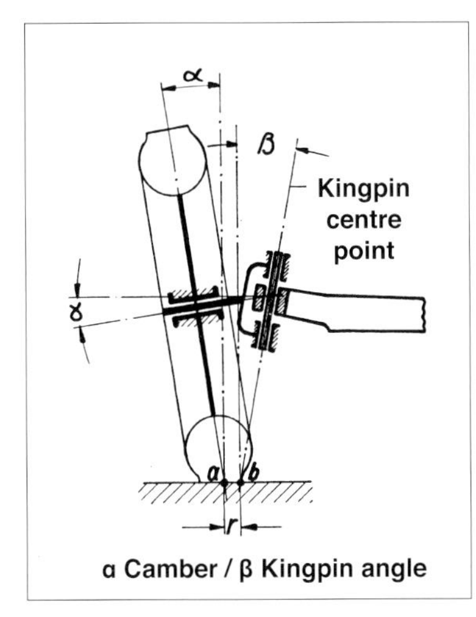

Positive camber and kingpin angle reduce the steering effort required, which was often an issue with prewar cars with solid axles.

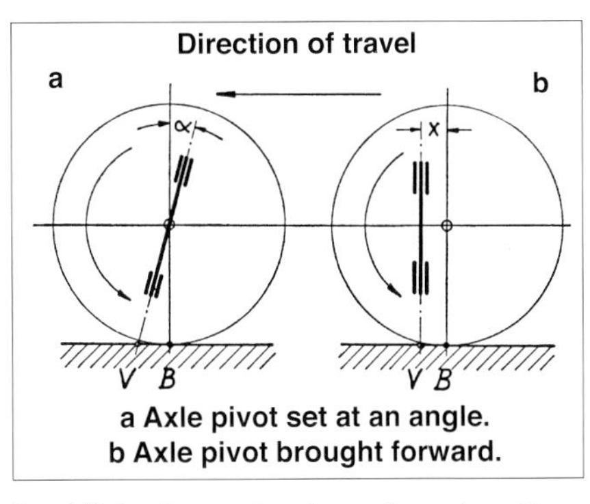

A stabilising degree of castor can be produced in several ways. In version a, the wheel leans into the corner.

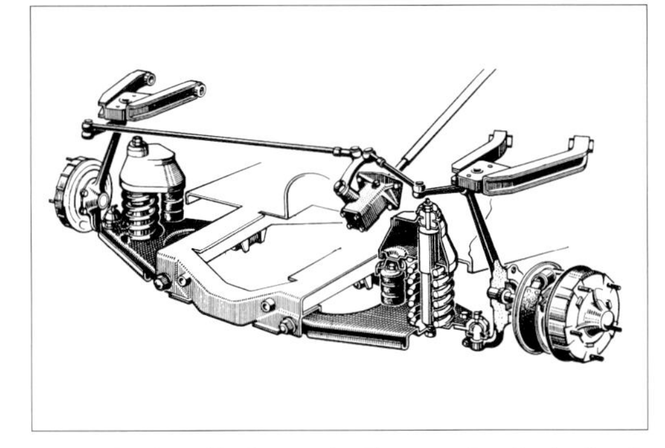

Expensive, but effective: on the Glas 1304, longitudinal and transverse arms control the front wheels.

The multi-link independent rear suspension (Mercedes 190 E) enabled the wheel to be perfectly positioned, regardless of the load it was subjected to. But it was extremely costly: up to six links were used for each wheel.

opposite is also true: all the components of a car's suspension should be as light as possible. A perfect example of how this can be achieved is provided by the rear axle of the Lancia Aurelia from 1950. In order to distribute the weight evenly, the gearbox, clutch and differential were all located at the rear and were bolted to the chassis, where these heavy components no longer contributed to the unsprung weight of the car. Even the brakes were mounted inboard. The wheels were individually suspended from semi-trailing arms: a lightweight design which also made it possible to adjust the camber of each wheel. This X-shaped design, with the trailing arm pivoted at an inclined angle, takes into account the strong lateral forces to which the tyres are subjected, which are unavoidable when cornering hard.

An NSU TT wouldn't be complete without negative camber. The X-shaped design made sure that the tyres had the best possible contact when cornering.

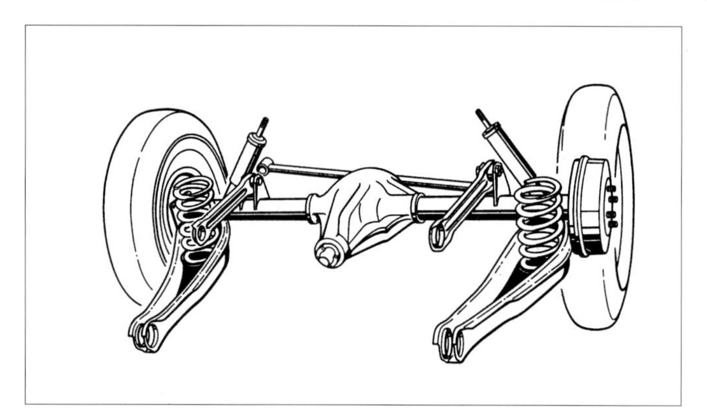

A hint of Ferrari: the live rear axle of the Opel Rekord was also located with four longitudinal control arms and a Panhard rod.

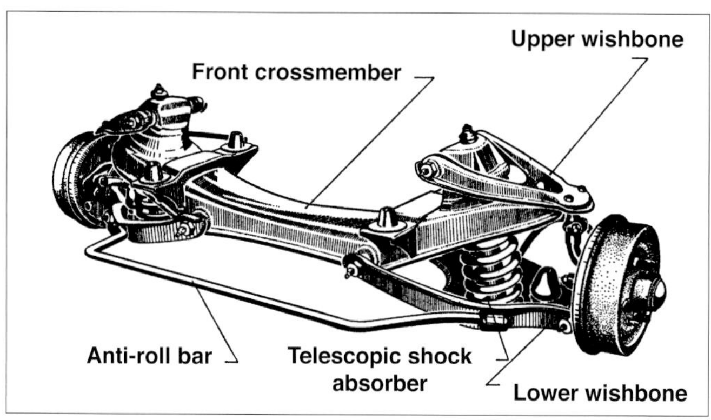

Opel once again: the double-wishbone front suspension is mounted on a separate crossmember to improve sound isolation. An anti-roll bar connects the lower wishbones to each other and to the body. In this way, roll when cornering is reduced.

Lancia's engineers put three basic principles into practice in this application. First, heavy components that are not part of the car's suspension should be kept separate from it. Experts make the distinction here between sprung and unsprung weight. The body and any parts attached to it contribute to the sprung weight, while the suspension and the wheels themselves are part of the unsprung weight. Second, in order to distribute the weight of the vehicle as evenly as possible across all four wheels, it's worth transferring heavy components in front-engined cars to the rear. Third, the means by which the wheels are located needs to be suited to the loads to which they will be subjected. The semi-trailing arm rear suspension proved so successful that Lancia's model was copied by several prominent companies, including BMW (for the 600, 700 and 'New Class' saloons) and Mercedes (with its 'Stroke Eight,' R107 and W116).

For their part, in 1982 Mercedes' engineers introduced a revolutionary innovation on the 190 E: the multi-link independent rear suspension. With this design, the movement of the wheel was controlled by up to six individual control links, each of which could move so that the wheel was always perfectly located – regardless of the load it was subjected to – including the camber and toe-in adjustment. It really came very close to an active steering system for the rear wheels.

There was a simple reason for the particular attention which the manufacturers paid to the rear axle in terms of their cars' roadholding. Each movement of the rear wheels has a noticeably more pronounced effect on the car's handling than that of the front wheels. And although the weight-related problems of traditional live axles were common knowledge, they became the most frequently used design in that era of car manufacturing when rear-wheel drive was the norm. That was above all due to the fact that the rear axle, differential and driveshafts could be combined in a single robust assembly which required little maintenance. Not for nothing did these supposedly primitive live-axled cars dominate the toughest rallies in the world for several decades, even though the inertia caused by this heavy design particularly penalised them on the bumpy tracks which the rallies followed. But while more sophisticated cars dropped out with broken

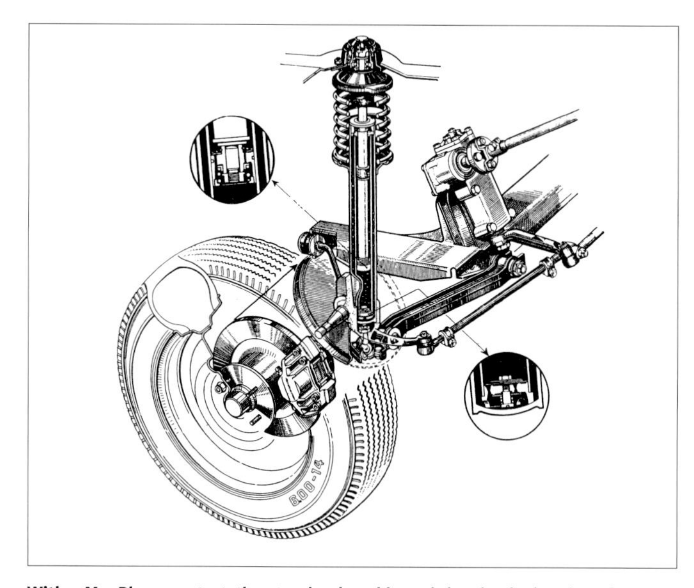

With a MacPherson strut, the steering knuckle and the shock absorbers form a single unit. The upper section, complete with the coil spring, is mounted so that it can pivot, while a wishbone controls the movement of the wheel (BMW 'New Class').

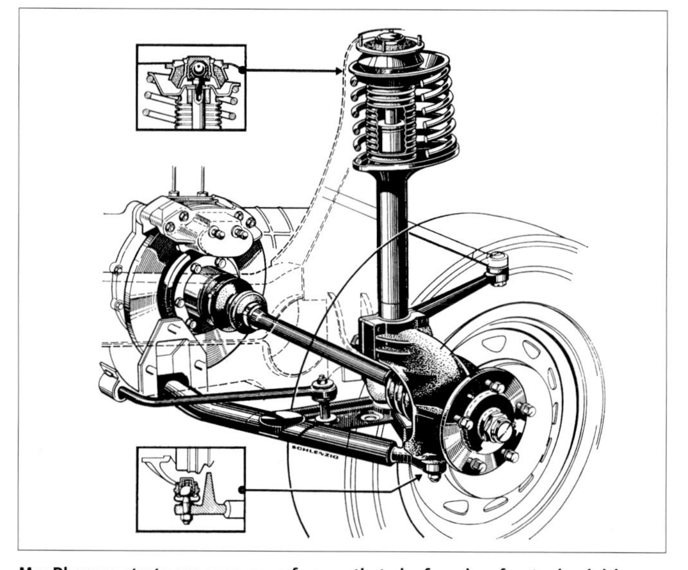

MacPherson struts are even more frequently to be found on front-wheel drive cars (in this case an NSU Ro80). The limited space required makes this compact design particularly interesting.

When a car 'cocks a leg' in a corner, the anti-roll bar shows it is having an effect. It transfers the spring force from the inner to the outer wheel.

driveshafts, the models with live axles reached the finish. In addition, these live-axled cars could be driven very precisely, provided they were fitted with trailing arms and at least one transverse arm (a Panhard rod).

A particularly interesting variation of the live rear axle suspension is the De Dion axle design, in which the heavy rear differential is bolted to the chassis and the axle consists only of a lightweight tube. This design was used, for example, on the big Opels of the 1970s and '80s, or on Alfa Romeo's Alfetta range (the latter with a rear transaxle).

For the steered front wheels, the solid axle lasted for much less time as the standard design in automotive manufacturing, with the exception of goods vehicles and all-wheel drive cross-country vehicles. The reason is obvious: whereas the rear wheels primarily need to run exactly straight ahead, the front wheels must respond to the driver's steering commands as faithfully as possible and without excessive effort. Since the wheel which is steered has to swivel about a kingpin (or individual supporting links), it is subject to forces when braking, accelerating or simply overcoming rolling resistance. The

Oversteer

Understeer

Whether a car understeers or oversteers depends above all on the position of its centre of gravity.

further the mid-point of the tyre contact patch is from the swivel axis of the steering knuckle, the greater the leverage forces exerted. Specialists also use the term 'kingpin offset' to describe the length of this lever.

In order to keep the steering effort as low as possible, the car manufacturers positioned the kingpin at an angle to the solid axle. This angle, with the topmost end of the axis of rotation slanted inwards, meant that the lower pivot point moved outwards. At the same time, they set up the wheel with positive camber, with the result that the tyre contact patch moved inwards and the kingpin offset was reduced. As long as the mid-point of the tyre contact patch lies outside the pivot point of the steering knuckle, this can be described as positive kingpin offset; if, however, it lies inside the pivot point, the kingpin offset is negative. The positive camber typical of prewar cars was a concession to the steering effort required, since the offset of the forged solid axle was fixed. It is easy to understand that the engineers did not want to put up with this stopgap arrangement for long, especially as it had another disadvantage. The outwardly-curved positive camber caused the front wheels to point away from the centre line of the vehicle, a tendency which could only be overcome by setting up the tracking so that the wheels were brought back towards each other. The specialist describes this as 'toe-in'. Tyre salesmen like this design, as the conflicting camber and toe-in settings are accompanied by heavy tyre wear.

In order that the wheels will return by themselves to the straight-ahead position when leaving a corner and then run as smoothly as possible, the upper end of the kingpin is inclined slightly rearwards, with the result that the lower pivot point within the tyre contact patch is shifted a couple of centimetres further forward. The wheel will now be pulled along – a principle which ensures that every supermarket shopping trolley can be controlled in a stable manner. Experts refer to this as 'castor.'

Unlike solid axles, independent suspension systems can be so designed that the desired effects only occur under specific load conditions. The very widespread variant of this with dual wishbones of different lengths, which are also used in the rear suspension of sporting cars, makes it possible to adjust the camber from positive to negative during the spring compression, which is ideal in terms of good handling. Even the castor can be set up so that it can be changed by means of easily adjusted mounting points for the transverse arm. As long as the lower transverse arm is exactly horizontal, the roll centre lies at the level of the road surface. The length and the mounting points of the transverse arms to the chassis frame or monocoque play a decisive role here. They determine the radius through which the outer supporting links will swivel and with it the complete set-up of camber, track, castor and kingpin offset.

A further reason for the quicker take-up of independent suspension at the front rather than the rear is the presence of the engine and all its accessories, which leave little room for elaborate suspension designs. In the interest of keeping the centre of gravity as low as possible, the engine and transmission are installed as far down as possible, while retaining adequate ground clearance. The suspension must make do with the remaining space on the left and right of the engine compartment, and few designs are more effective in this respect than a MacPherson strut.

In this instance, the upper mounting point – which can pivot – is attached to a compact unit comprising the shock absorber and coil spring, while the bottom end serves as the hub carrier and forms a single unit with the steering knuckle. A wishbone, which should be as long as possible, absorbs the forces transmitted by the wheel at its bottom end. The potential disadvantage of this design – that the track width is reduced during the spring compression (or if the suspension is lowered) – is more than offset by its compact construction. It is no surprise then that the MacPherson strut has become the standard design for front-wheel drive cars with transverse-mounted engines. Meanwhile, it is also increasingly found – minus the steering components – on the rear axle.

A true classic: in this instance, the leaf springs take over the job of locating the wheels. Unwanted lateral movements and reactions when coming on and off the power are the result.

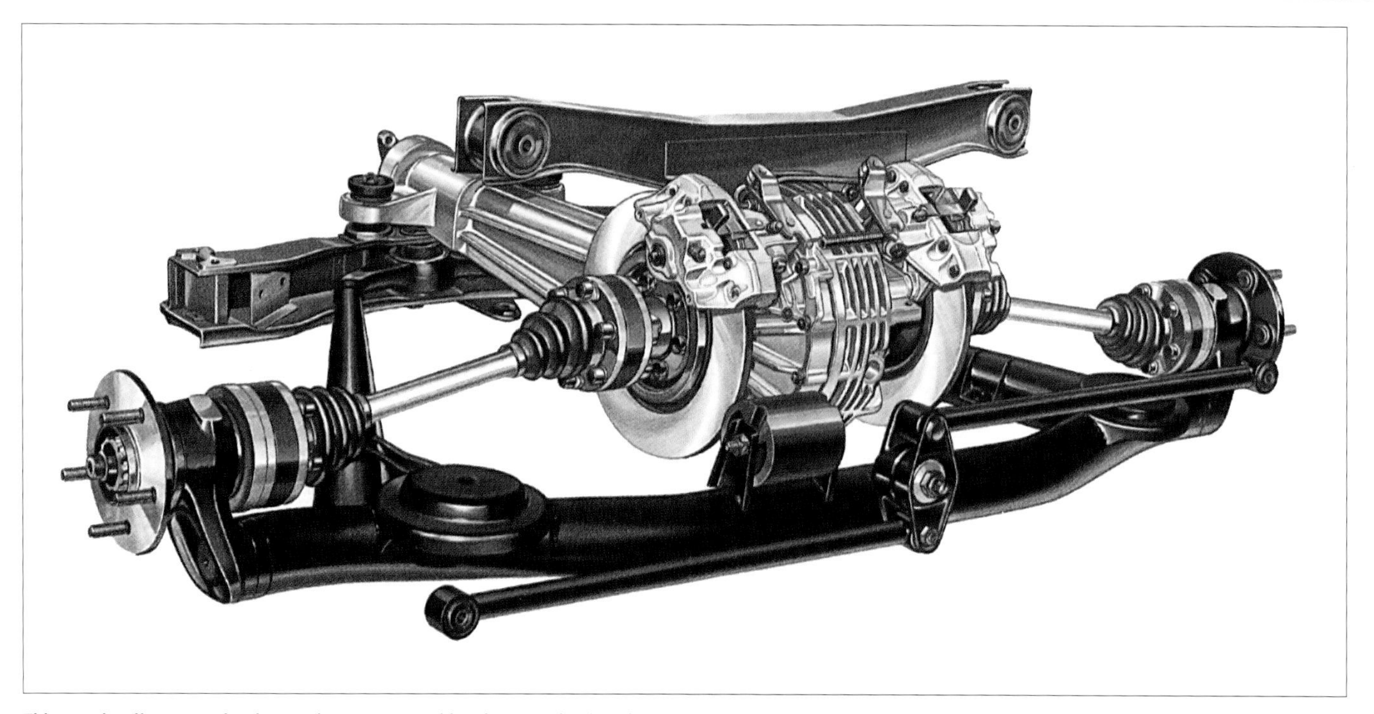

This, too, is a live rear axle: the De Dion system enables the rear wheels to be controlled with a constant degree of camber, and without the weight penalty of a conventional live axle. In the Alfa Romeo 6 shown here, the lateral location is assured by a Watt's linkage.

The anti-roll bar plays a supporting role alongside the springs and other suspension components. In design terms, it is a U-shaped torsion-bar spring, which is attached by means of a rotating mount to the car body, and the ends of which are connected to the suspension at either extremity of the same axle. If both springs compress together – as when driving over a bump – the anti-roll bar has no effect. If, however, the body of the car leans to one side in a corner, the anti-roll bar transfers the compression movement of the outer wheel to the unloaded inner wheel.

In the case of very firm sports anti-roll bars, this effect can be so pronounced that the inner wheel lifts off the ground when cornering. The roll centre of the axle cannot be changed using an anti-roll bar, the effect is more akin to a transfer of the spring stiffness from the unloaded to the loaded side of the car.

Speaking of springs, in chassis construction leaf springs are quite unjustly considered as an antiquated solution from the days of the stagecoach. That certainly applies to the simple live axle, made up solely of two sets of leaf springs and a pair of shock absorbers – as was for long the custom with American cars. Considered solely as a spring, a leaf spring is not necessarily inferior to a coil spring or a torsion bar. The modern, sporty chassis of the latest Corvette models prove that to great effect. If, however, the forces of acceleration, braking and cornering on the live axle are not alleviated by the fitment of compression struts or transverse arms (a Panhard rod or Watt's linkage), the suspension characteristics will vary according to how much pressure is applied to the accelerator or brake pedal, not to mention the lateral movement when taking a corner. The reason for this lies in the changed friction which occurs between the leaves, which can literally become wedged together when a powerful American V8 sends all its torque to the rear wheels. In principle, therefore, in good suspension designs, each component should have its own distinct task. Springs take care of springing, shock absorbers absorb shocks, and the wheel carriers locate the wheels ... but you most likely worked that out for yourself already!

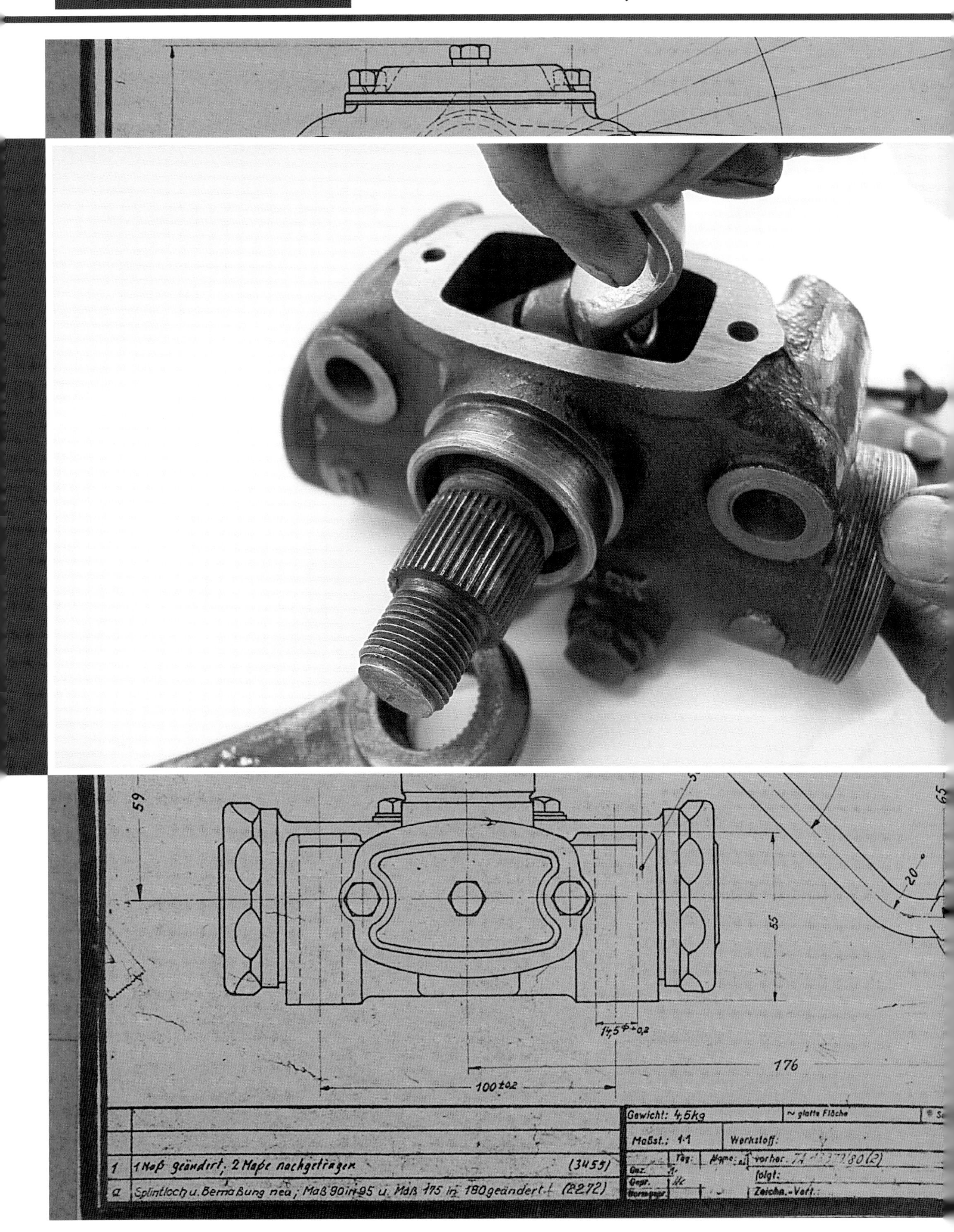
59
55
176
100±0,2
Gewicht: 4,5kg
~ glatte Fläche
Maßst.: 1:1
Werkstoff:
1 1 Maß geändert; 2 Maße nachgetragen (3455)
a Splintloch u. Bemaßung neu; Maß 90 in 95 u. Maß 175 in 180 geändert (2272)
Tag:
Name:
vorher.
folgt:
Gez.
Gepr.
Zeichn.-Vert.:

Shock absorbers
Friction and lever-type models

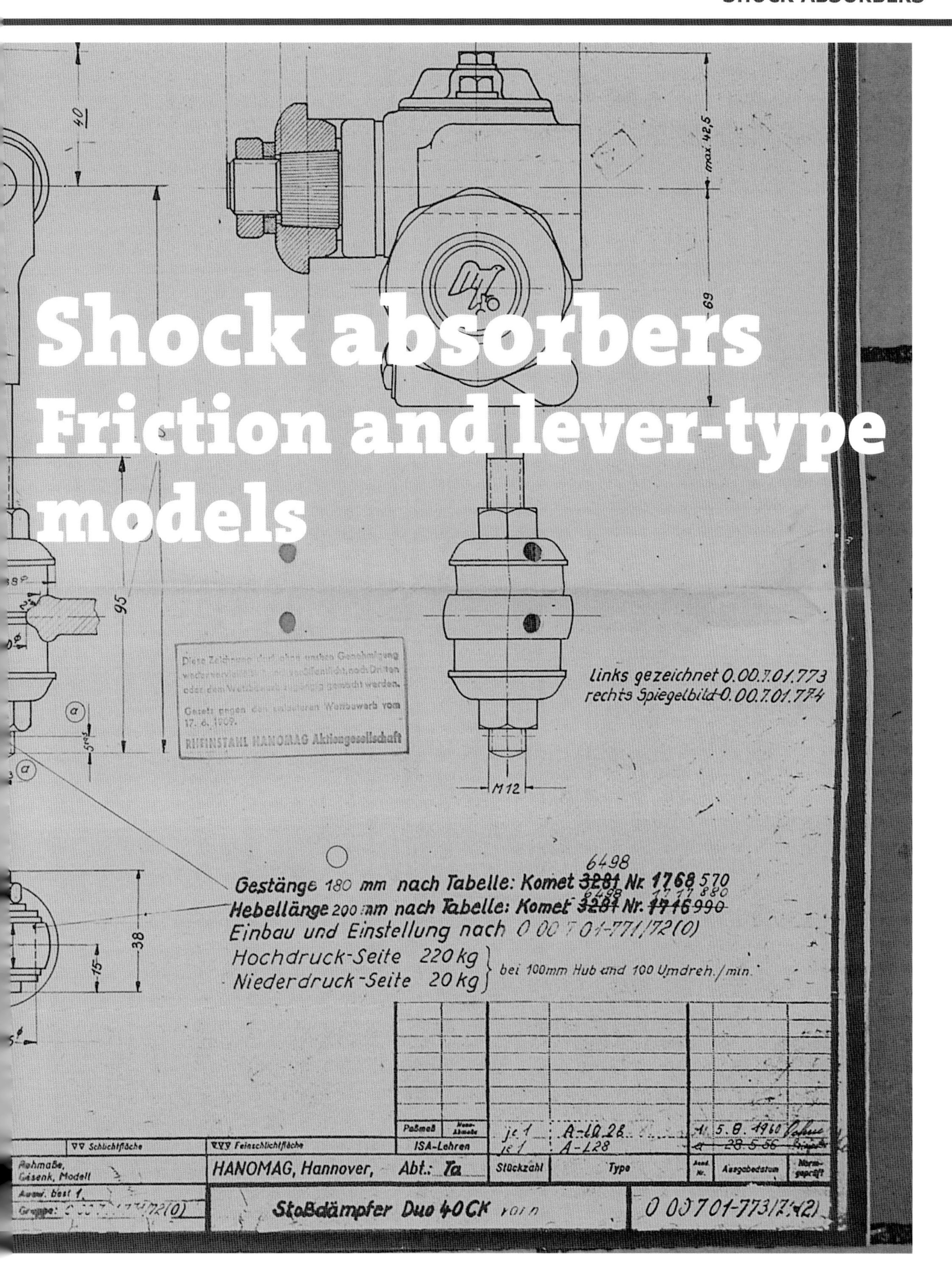

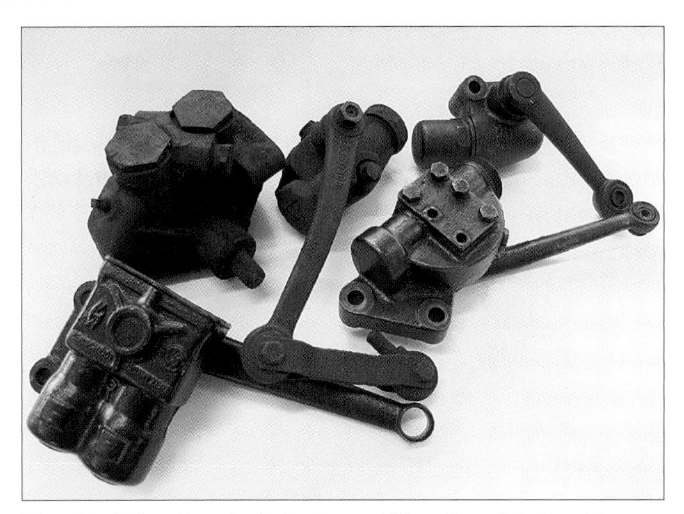

A broad topic: lever-type shock absorbers exist in a wide variety of models.

Without the calming effect of its shock absorbers, a car's suspension would constantly dance a jig – with dangerous consequences for the car's roadholding and handling! In this and the following chapter we take a closer look at three popular models of shock absorber: the friction, lever and telescopic types.

Left to its own devices, a car's steel suspension would simply jump up in the air when going over every pothole. This is where shock absorbers come in, which have helped settle cars' suspension for nearly 100 years.

A shock absorber is really a vibration damper. It absorbs part of the energy which occurs alternately as spring tension or as movement of the car's bodywork. The energy which is removed from the car as it vibrates is converted by the shock absorbers into heat. Well-adjusted shock absorbers do this in such a way that, on the one hand, a considerable degree of suspension travel is used, and few shocks are transmitted to the car's body; and, on the other hand, the body does not go on vibrating after a single compression and rebound cycle.

When motoring was in its infancy, the damping required was provided by the intrinsic damping of the leaf springs themselves, with their leaves rubbing against each other. With the breakthrough of pneumatic tyres and increasing speeds, however, the damping provided by the springs was no longer sufficient.

This was when the scissor-type friction shock absorber came into being. In this design, one of the 'blades' of the scissors is connected to the chassis frame or body, the other to the axle or wheel carrier. Where the blades pivot, there are spring-loaded friction plates made from metal, wood, leather or other materials, as were also used for clutch and brake linings. Between them the kinetic energy of the car's bodywork is, to a certain extent, dissipated as heat.

From the early 1930s or thereabouts, hydraulic lever-type shock absorbers gradually replaced the friction-type shock absorbers which had been popular until then. The latter survived for a few more years in the world of motorsport, as they could be quickly and easily adjusted, while lever-type shock absorbers became dominant for mass-produced cars, and were only seen off in the 1950s and 1960s by the telescopic shock absorbers which remain current today.

The operating principle of the hydraulic shock absorbers still in use is always the same, regardless of their exact configuration: the vibrational energy acts on a piston

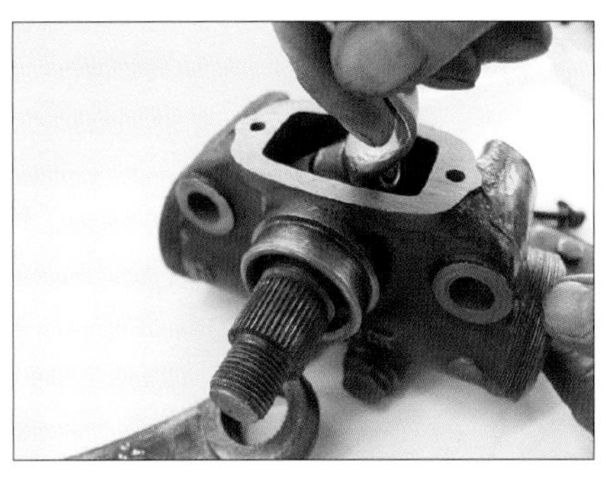

A job for specialists: the outer lever has already been pulled out. After releasing ...

... the screw, a puller can now be used to extract the rod and remove the piston lever.

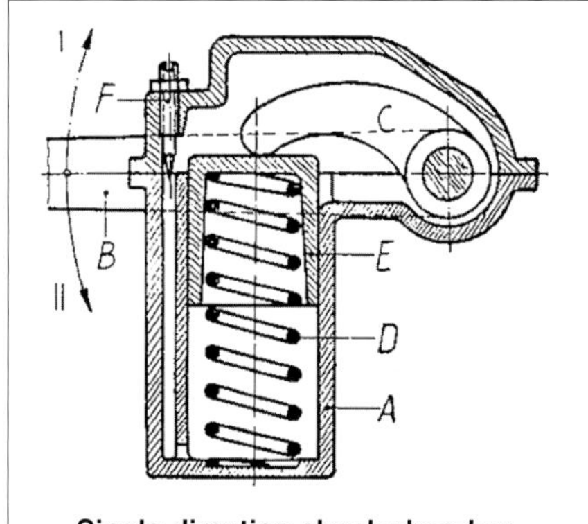

Mono-type lever shock absorber: only the extension or 'rebound' stage is hydraulically damped.

Control valves: after the sealing caps have been taken apart, the valves ...

... can only be unscrewed with a special tool (which can, however, be easily fabricated).

An easily understood design: with the right tools, the shock absorber can be quickly stripped down into its two dozen or so parts. Now it's time to take stock ...

inside an hydraulic cylinder, which, in the course of its movement, forces hydraulic fluid through narrow bores or valves. It is this resistance to the flow of the fluid that reduces the spring energy.

The effect of the shock absorber is, therefore, dependent on the speed at which it operates: the faster the fluid is pushed through the openings, the more force is required.

The lever-type shock absorber works in the following way: the housing of the shock absorber is bolted to the chassis, while the lever (which gives this design its name) is joined to the suspension (or, as on the MGB, actually constitutes the upper transverse arm). This lever is positioned on a rod, at the inner end of which the piston lever is located. The piston lever has a cam which protrudes into an opening of the shock absorber piston. By means of two small pressure plates made of hardened steel, the piston lever cam transfers the movements of

A precision job: on this machine from the 1930s, the fine gear teeth are cut on blank rods. This supplier manufactures all the components himself, right down to the housing (left).

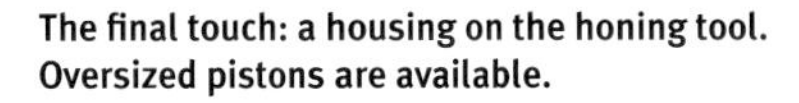

The final touch: a housing on the honing tool. Oversized pistons are available.

Cleaning the teeth: on this machine, the inner gear teeth of the piston lever are reworked.

the suspension to the piston, and causes it to move in its oil-filled cylinder. In this way, the hydraulic fluid is forced through the 'control valve' which regulates the shock absorber.

Lever-type shock absorbers are very long-lived, but after 40 or more years, problems can arise. Amateur mechanics, however, have only limited options. These come down to regularly checking the fluid level, which should reach about 5mm under the upper edge of the housing. A regular oil change every couple of years also makes sense, to get rid of dirt and metal particles. After changing the oil, the level should be checked after driving 50km (30 miles) and topped up as necessary.

Experienced mechanics can also influence how the shock absorbers are set up, by adjusting the control valve. These are spring-loaded screws which determine the cross-section of the channel through which the fluid passes. The larger this cross-section, the softer the damping; the tighter the screws are turned down, the firmer the damping. The car's responsiveness can, in turn,

In addition, the levers now have a precisely defined cam, which the supplier brings back into shape using a special machine.

The death sentence: if the opening for the rod has widened and become oval, an overhaul is no longer worthwhile.

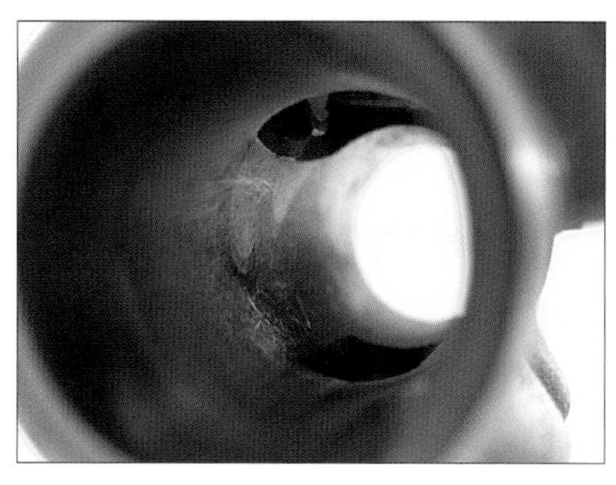

The cylinder: rust or run marks here don't mean it's a disaster beyond repair.

Old metal: a corroded shock absorber rod with a groove which has been 'milled' by the dust cap.

Piston lever: the cam shows pronounced run marks; the lever will have to be replaced.

be controlled by the springs, with stiffer springs achieving the same response.

Special tools are needed for these jobs, but you can make them yourself. The downside: it can be hard to find precise setting data – 'trial and error' is the rule here.

It is easier to alter the setting by changing the viscosity of the oil. Pressing thicker oil through the holes takes more force than for a thinner oil, and using thicker oil makes the shock absorber firmer. Shock absorber oil is not sold everywhere, but you should have a good chance finding it in specialist motorcycle shops, where it is sold as fork oil, or in larger/better stocked spare parts shops.

Adjusting shock absorbers isn't rocket science. To do so, you remove the shock absorber and attach a weight to it, then measure with a stopwatch the time it takes to go once from end to end, measuring the compression and rebound cycles. By filling it with new oil, you can achieve your desired setting: either a longer time (for a firmer set-up) or a shorter one (for a softer set-up). If you have stocked up in advance with both a thinner and thicker oil, you can blend the two together until you reach the desired setting. In this case, it makes sense to use oils from the same manufacturer and product range. And finally, it is important that the shock absorbers on the same axle should be calibrated in the same way!

Apart from maintenance and adjustment jobs such as these, the opportunities for amateur mechanics are very limited. If fluid seeps out of the housing cover, it is easy to cut a new joint to length, but leaks from the shock absorber rod usually present more significant problems for restorers. This is because the felt washer behind the lever is defective and is almost impossible to replace, as the lever is firmly pressed onto the rod and can only be released using a special puller.

A similar problem applies when there is free play between the connecting parts inside the shock absorber. This can only be checked once the lever has been separated from the rod and can move freely. Since lever-type shock absorbers were produced in large quantities and remain available today for an amazingly wide range of cars, a repair can only be justified in exceptional cases. New units or reconditioned exchange parts can be found at dealers specialising in that make of car. Clubs and online communities can often be of additional help.

If you can't find the part you are looking for to replace a faulty shock absorber, you will need a really well-equipped workshop. If a shock absorber has become completely dry, you will generally need to repair the bearing surfaces. But since oversized pistons are often available, drilling and honing the bores is not the answer. In this case, a cylinder liner must be inserted. Making new bearings, on the other hand, is an easy job. Most owners, however, switch to shock absorbers which are still available and adapt these as necessary.

In place of a felt ring, the housing is milled to take a sealing ring.

One for specialists: a newly fabricated rod (and lever) with a reference gauge for the gear teeth.

A maximum of 0.5mm (0.02in) free play: a piston lever with pressure plates cut to size.

To see just how time-consuming the overhaul of a popular Fichtel & Sachs shock absorber really is, we looked at a specialist company based in Würzburg. This traditional business, founded in 1932, acquired licences from Fichtel & Sachs in the 1950s, and since then has produced several types of shock absorber, which are found, among other applications, in classic Mercedes (such as the 'Adenauer' and Unimog models). The expert on these is the boss himself.

"In general, an overhaul is no longer worthwhile if the inner workings of the shock absorber have corroded due to a lengthy shortage of fluid," the expert explains. He sets the upper limit for repair costs at about ●x310; if the overhaul will cost more than this, he recommends fitting a new replacement for about ●x620. Even when the fluid level is correct, though, a shock absorber can be worn out, which will only become apparent after it has been stripped down.

First of all, the specialist removes the lever from the shock absorber using a puller, after cleaning the exterior and draining the oil from it. Next, after the top of the housing has been taken apart, the piston lever can be accessed through it and unscrewed; then, again using a puller, the rod can be extracted. The first signs of damage are often evident here, such as rust marks on the bearing surfaces, corroded and damaged fine gear teeth, or traces of dust caps which have worked their way into the surface of the rod. Generally speaking, the felt gasket will have had it, but a more serious problem occurs when the housing has widened into an oval shape where the rod passes through it. If everything is alright, the specialist will open up the housing to replace the felt with a sealing ring.

The piston lever can also now be fished out of the housing. It typically shows signs of wear on the cam, which is subjected to high pressures when operating: no wonder, given that several centimetres of suspension travel at the wheel are reduced to a stroke of just a few millimetres here.

The unspolit romance of industrial machinery: as the final stage in the job, the overhauled shock absorbers are adjusted on this test bench. The load applied is measured and can be read off on the scale at the top.

The cam has a defined shape, and for new parts this can be obtained using a special grinding machine. The small hardened metal plates, between which the lever cams are located with just 0.5mm free play, are also liable to wear.

After a number of larger and smaller caps have been disassembled, the piston and control valve can be removed, the latter using a special lever. "For the most

Using the resources to hand: the shock absorber can be tuned by means of the oil's viscosity.

Apart from changing a gasket, there's little more an amateur mechanic can do.

part, the cylinder must be honed, and an oversized piston fitted, so as not to exceed the tolerance of two hundredths of a mm," the owner of the business explains.

It takes a good many hours' expert labour before a shock absorber – with new rods, pistons and piston levers, and complete with its pressure plates, valves, gaskets and sealing rings – can be reassembled, refilled and adjusted on a special test bench.

Rebuilding a lever-type shock absorber takes even more effort. The Würzburg-based business only buys in

Friction-type shock absorbers today are straightforward to restore: in contrast to hydraulic systems, they suffer less deterioration when not in use, and are relatively durable when fitted with modern friction linings.

Friction disc shock absorbers

Ultimately, the durability of these shock absorbers is far from ideal. With friction linings made from natural materials such as leather, cork or beech wood, often it only takes the wear caused by a single journey to ruin the previous adjustment of the shock absorbers. In the 1920s, Lancia overcame this shortcoming with friction dampers which could be adjusted hydraulically from the driver's seat. In any event, the constant need to make adjustments when the car was in its infancy and cylinder heads had to be removed and decoked every few thousand miles, attracted little adverse comment. Nowadays, however, even drivers of prewar cars place higher demands on their vehicles in terms of their durability and reliability.

Modern friction discs live up to these higher expectations, using the same materials as clutch and brake linings. "If you really want to drive your car, you should definitely turn to these discs. Wood, cork and so on are better suited to show cars," advises the well-known restorer Peter Bazille from the B & F Touring Garage in Troisdorf.

If the right setting cannot be found, despite fitting modern friction discs, it may be that the disc spring assembly has lost tension. This suspicion is confirmed the die-cast housings; all the other components are made up in its own workshop, many of them on machines which have been 'classics' in their own right for many decades, something which certainly doesn't harm the quality of the work they do. "I have one customer from the USA who comes to see me every two years and fills two suitcases with shock absorbers for his Mercedes 300, then flies home again," says the shock absorber expert, who currently has only one problem: he cannot find a successor who is interested in the subject of lever-type shock absorbers.

Whatever you need: replacement parts for a Hartford shock absorber (at the back, a modern friction lining).

if the disc springs hardly stand upright on a flat surface once they have been removed. In this case, the disc spring assembly will need to be replaced.

"The springs, friction discs and setting scales for common shock absorbers like the Hartford are all available new," explains the prewar specialist Jochen Schramm from the firm Medidenta Schramm.

Apart from the components of the friction assembly, wear otherwise occurs only on the mountings for the transverse arms. On many cars, the retaining bolts can be changed; alternatively, bushes will need to be made up.

The final stage is to adjust the newly overhauled shock absorbers. The scale underneath the disc spring provides an indication. With the help of this, all four shock absorbers are set to their softest position. This initial rough-and-ready adjustment is followed by a practical test. Using a ruler, the distance is measured between a distinctive point such as the bumper and the ground. If you now press down on the corner of the car which is being measured, the shock absorber should be adjusted so that the body of the car returns smoothly to its original position without bouncing. In the same way, the rebound motion should not 'stick' so that the assembly no longer springs back to its original position. This test is then repeated on all four wheels, and you are done! The beauty of all this is that the repair jobs once carried out by a village blacksmith will not be too much for a restorer either.

Clearly seen: the mounting points for the shock absorbers produce wear on the friction discs.

On a smooth surface, the springs should not be flattened, but stand clearly upright.

BILSTEIN
BILSTEIN
BILSTEIN
F4-BMS

Shock absorbers
Choosing and fitting the right shock absorbers

Sports shock absorbers are the key to improved roadholding. The firmness of the best models can be adjusted to suit the driver's taste – additional factors, though, determine the true character of the shock absorbers produced by Bilstein Tuning – and to match the 'seat-of-the-pants' feel of an experienced test driver.

"You can learn the seat-of-the-pants feel," joked the head of the Bilstein test driving department, which is responsible for setting up shock absorbers for specific models of car day after day, "but it takes some time!" He should know. The 'seat-of-the pants feel' refers to the subjective sense of roadholding, handling characteristics and safety, which highly experienced drivers develop; an infallible and important metric for manufacturers. Despite this, and right at the start of our discussion, the test team leader quickly put paid to one piece of wishful thinking: "There is no perfect set-up, at least not for cars used every day. This can be put down to the variety of road surfaces and load conditions with which production cars are confronted. Setting up a competition car for the smooth surface of a racetrack, on the other hand, is almost child's play!" And so the expert turned our view of the world on its head: until now, we had thought that competition chassis were the ultimate test for every shock absorber specialist.

With racing cars, their weight – apart from the contents of the fuel tank – remains virtually constant. A powerful saloon car, on the other hand, may have to carry five passengers and their luggage one day, but only the driver the next. In the morning, it may be speeding along a freeway, then in the evening head off along a potholed track for the weekend.

The shock absorbers must be firm enough that they do not bottom out when the car is fully loaded, but without rattling the passengers' fillings loose. The Bilstein test team leader and his colleagues simulate all these extremes on a test track at Papenburg in northern Germany, where consistent test cycles can always be repeated. In this way, they can learn how even the slightest adjustment to the shock absorber settings can affect each particular car.

"The same car is always driven by several testers one after another, so that afterwards we can discuss the pros and cons of each new set-up. In theory, the behaviour of a shock absorber can, of course, be assessed on the test bench. At Bilstein, however, we allow ourselves the added luxury of driving tests, as the driver's 'seat-of-the-pants' feel is more precise. Besides, the car will later be driven by a human being," the specialist added.

The test drivers need special expertise above all to estimate correctly the so-called 'velocity' of the shock absorber. This term signifies the speed at which the piston working inside the shock absorber tube moves. With slower movements, such as when braking or cornering moderately, other components inside the shock absorber respond than when it undergoes hard braking or drives over a pothole. "If the rear of the car rises sharply under braking, there are two ways to prevent this. Either you set up the rebound stage of the rear shock absorbers to be somewhat firmer, or you adjust the compression stage of the front shock absorbers to be firmer. Knowing which

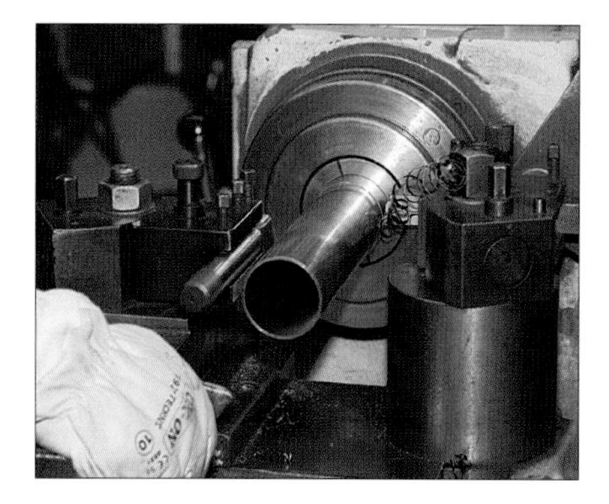

The length is what counts: for lowered suspensions, the shock absorber tubes are cut to fit.

With a million 'limited-production' shock absorbers produced each year, that makes for quite a few metal cuttings left over.

First, the tubes are sealed and welded to the mounting ring.

The week and year of production are engraved on the tube so that they can be clearly read.

Character-building: the working pistons and spring washers later determine ...

... how the shock absorber will respond. At the top is the 'piston ring' made from a special plastic.

Take care! This nut can work loose if an impact wrench is used on it.

The shock absorber can only be opened and filled with fluid using a special press.

Here, a one-off model is being filled with nitrogen by hand.

The first 'test drive' takes place on the test bench. A computer records its characteristics.

The sports shock absorbers can be recognised by Bilstein's typical yellow colour and blue stripes.

An exception: the light-alloy components of these adjustable coilover units remain unpainted.

is the right course to take depends on the how the car handles otherwise.

"In any case, it makes sense to fit shock absorbers at the front and rear which have been set up to work together. For this reason, in our catalogue we always present the appropriate combination for each model of car," the expert explains. But what exactly happens when the specialists change the set-up? To understand this, let's take a look at the inner workings of a modern gas-filled shock absorber, and how it is produced. The German town of Ennepetal is home to Bilstein's head office, and to its tuning department, which every year produces a million shock absorbers in small batches. A million, in small batches? This isn't a contradiction, as this is where tuning firms, racing teams and owners of classic cars all come to have shock absorbers built which cannot be supplied off the shelf, or which are no longer available. The remaining eight million shock absorbers every year are produced in large runs, which Bilstein supplies on an OEM basis to Porsche, Mercedes, BMW and other manufacturers.

In general, there are two different designs of hydraulic telescopic shock absorber. The twin-tube shock absorber has a valve at the bottom of the actual damper, through which the hydraulic fluid can flow into an outer tube. This is needed to offset the displacement of the fluid by the piston rod during the compression and rebound stages. Fluid flows through the working piston, at the end of which the rod is located, and thus ensures the actual damping action.

The mono-tube gas-filled shock absorber, which is the norm today, offsets the change of volume from the movement of the piston rod by the addition of a charge of gas, which is compressed as the piston rod enters the cylinder. This has several further advantages: whereas the twin-tube shock absorber must always be installed with one specific side to the top – the maximum angle is about 45 degrees from upright – the mono-tube shock absorber can be mounted in almost any position. At the same time, the gas deals with a typical problem of hydraulic shock absorbers, known as foaming. To understand this, it is important to realise that all hydraulic fluids contain a certain proportion of gas.

If, as the result of the movement of the piston, a vacuum is created on one side, the gas escapes and allows the

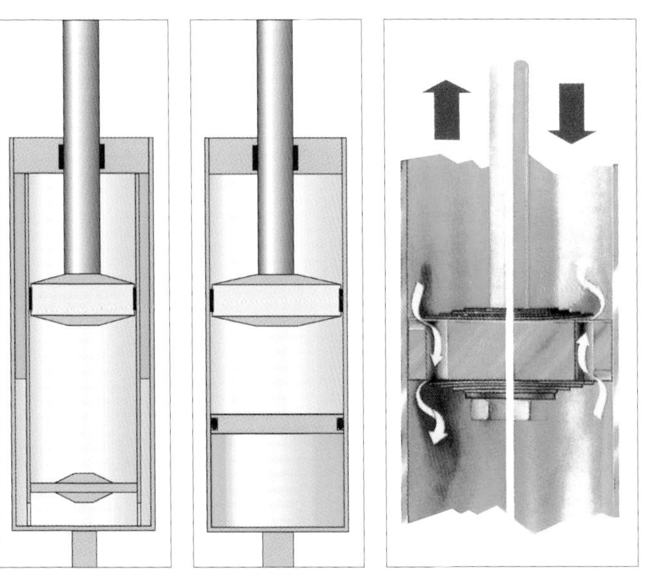
A cross-section of the twin-tube (left-hand image) and mono-tube shock absorbers: the flat floating piston separates the nitrogen from the fluid (middle image). As soon as the working piston starts to move, the control valves open (right-hand image).

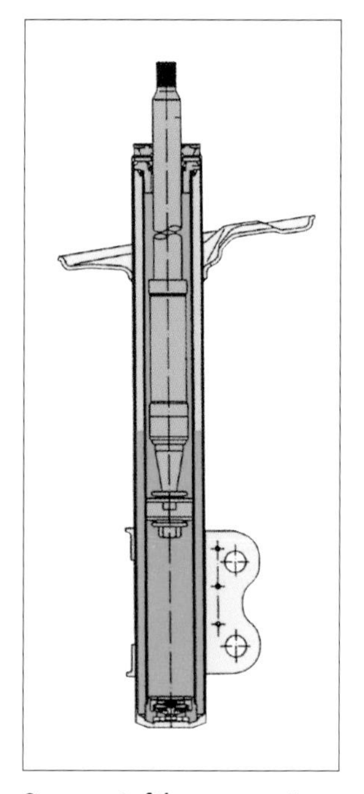

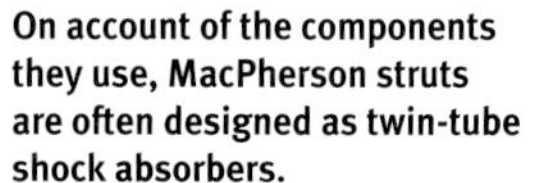

On account of the components they use, MacPherson struts are often designed as twin-tube shock absorbers.

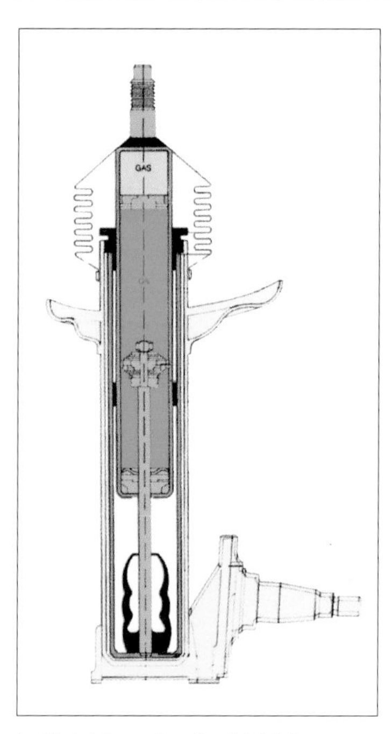

In Bilstein's version, the thick tube absorbs the forces produced in controlling the wheel.

The 42-page Classic catalogue lists Bilstein's shock absorbers from Alfa to Zastava (this is an older version, but it still exists today).

The glass demonstration shock absorber shows the effect of the gas pressure. On the left, the piston moves up and down inside the clear hydraulic fluid. In the absence of any pressure, the fluid foams up, due to cavitation.

fluid to foam. It is easier to imagine this effect by taking a bottle of sparkling mineral water as an example. As long as the bottle remains under pressure, the carbon dioxide remains dissolved in the water. If, however, you open the cap, the gas bubbles up to the surface of the water. If the gas does not dissolve at normal atmospheric pressure, but only when there is a vacuum, this effect is known as cavitation. The charge of nitrogen gas, which is held under a pressure of 360-500psi depending on the position of the piston rod, prevents the formation of foam and thus ensures that the damping effect of the fluid remains virtually constant.

Gas-filled shock absorbers are accordingly easy to recognise. In its unloaded state, the piston rod always moves out to its stop, with the result that the full length of the shock absorber can be seen. There are different versions of the mono-tube shock absorber, which are essentially distinguished in terms of how the gas is separated from the hydraulic fluid. Bilstein makes use of an idea of the French researcher into oscillations, Christian Bourcier de Carbon, and employs a floating piston, which is only guided between the walls of the shock absorber housing.

The key role, however, is played by the working piston at the end of the rod. The light-alloy casting has channels which are optimised for good flow, and which are sealed at the top (compression) and bottom (rebound) with flexible spring washers. As soon as the piston starts to move, the flow of fluid makes the spring washers move up. The response characteristics of the shock absorber can therefore be affected by using different strengths of spring washer. How each particular set of shock absorbers is assembled, on the other hand, depends on the outcome of the road tests.

As far as the limited-production series are concerned, the shock absorbers are assembled following a component diagram. On the one hand, this means that further fine-tuning is only possible after opening up the shock absorber; and on the other hand, it takes a good deal of insider knowledge to swap one particular spring washer for another. In practice, any attempts to do so usually get no further than the well-intentioned warning 'Do not open' on the sealing cap of every Bilstein shock absorber. If someone does, nonetheless, attempt to carry on without using the special tools required, they risk seeing the individual components shoot past their ears with a pressure of 360psi. If you are left in any doubt, that could prove fatal!

Here is what Bilstein's experts have to say: "In principle, customers can send us their shock absorbers so that we can adjust the set-up. To do this, we need exact specifications, which hardly any outsiders are able to provide. Consequently, the department we have set up specifically to cater for customer special orders tends to deal more with repair work and custom-made models. Practically any Bilstein shock absorber can be repaired

To remove the shock absorbers, the professional easily raises the lower transverse control arm of a Mercedes using an hydraulic gearbox lift. The shock absorber can then be released.

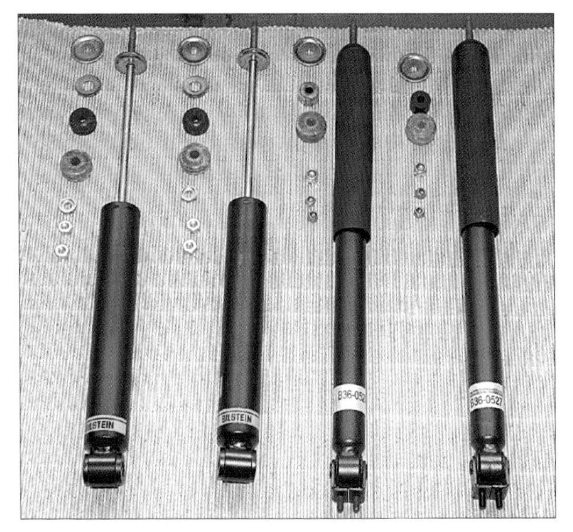

This is how the complete set of shock absorbers looks, laid out ready on the workbench.

To avoid breaking off the bolts, some rust remover is first sprayed onto them.

or rebuilt here, if the model in question was previously available, but is no longer in stock.

"Special orders can sometimes be rather more unusual. A few months ago, for example, we replaced the MacPherson struts on a 02-series BMW with an adjustable coilover set-up."

Does the tuning of the suspension, insofar as the shock absorbers are concerned, always need to be carried out by the factory? Bilstein, as well as other manufacturers such as Spax and Koni, also supply shock absorbers which customers can adjust themselves to suit their particular requirements. With the Spax units, it only takes the turn of a screw; with the Konis, the shock absorber is released on one side, pushed together and then adjusted. The more expensive Bilstein sports shock absorbers can even be adjusted by means of a knurled nut, with hardly any tools required.

Making these adjustments does not, however, change the characteristics of the shock absorbers, which are determined by their design, but rather their overall firmness. If you don't mind going to the trouble, you can set the rear shock absorbers to be somewhat firmer when the boot is fully loaded, or rather softer if you are out for a spin on a Sunday without the family.

In any case, firmer doesn't automatically mean better. This is because a fully tightened shock absorber will transmit more shocks to the car's body: on the one hand, this will cause more wear and metal fatigue, and on the other, it will make the wheels start to bounce on bumpy roads. Not only does this look silly, but, when driving on the limit, it is actually slower and may sometimes be dangerous.

Shock absorbers also play an important part when the suspension is lowered. Since the actual effect is achieved by fitting shorter springs, many tuning fans forget that the shock absorbers must be adapted to suit them. This starts with the fact that, on most axles, the shock absorbers limit the maximum possible rebound. Unless at least a rebound or spring travel limiter is fitted, it is possible that when a hump in the road is taken in too spirited a manner, the brand-new spring will simply go all the way through, and the car will subsequently bottom out completely.

The spring travel limiter is nothing more than a plastic buffer, which limits the maximum rebound so that the spring is always pre-tensioned.

In many cases, lowering the suspension requires a complete set of new shock absorbers, since, depending on where they are fitted, the so-called central position of the piston is moved massively upwards. This means that the shock absorbers reach their upper stop before the springs are fully compressed. In this case, Bilstein offers

After the lower mounting bolt has been undone, the fitter turns to the engine compartment.

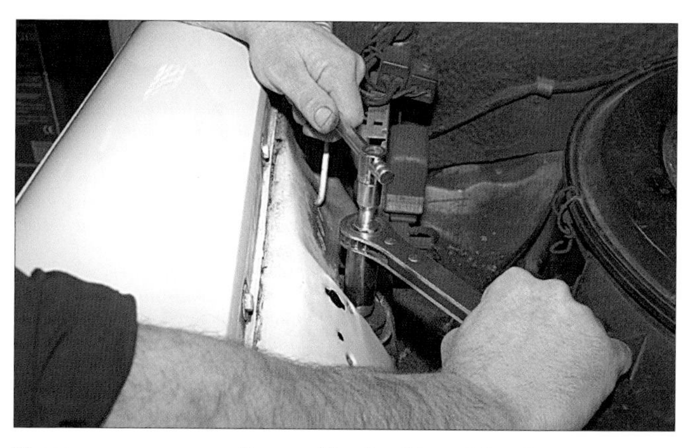

Please never use a pneumatic screwdriver here! A special ratchet spanner ...

... prevents the piston rod from turning at the same time as you undo the shock absorber.

The rubber bump-stop and the upper supporting plate can now easily be removed.

Clearly seen here: the top of the piston rod has been levelled off so that a tool can be attached to it.

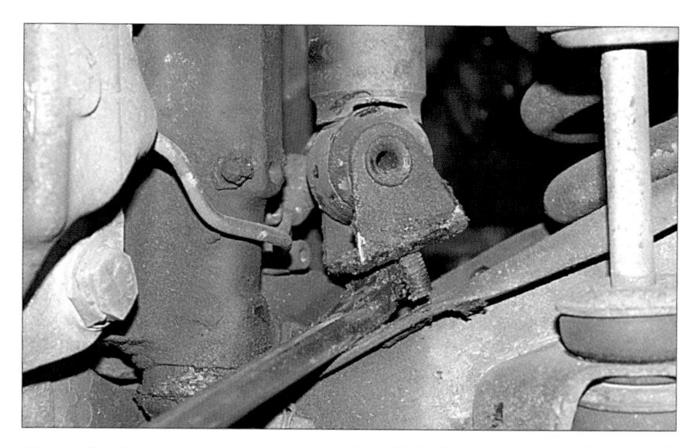

The mechanic now uses a tyre lever to prise off the lower mounting of the shock absorber.

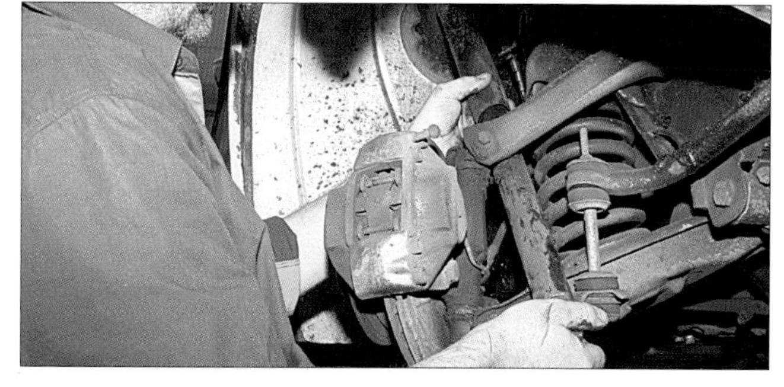

On the Mercedes there is enough room to pull out the shock absorber, despite the wheelarch liner.

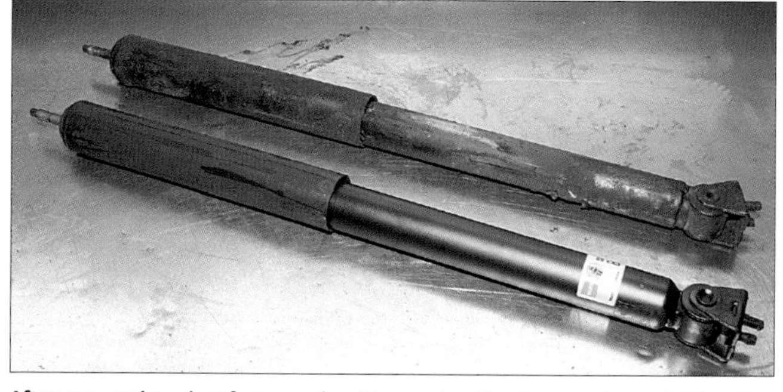

After 250,000km, time for an early retirement: a direct comparison of old and new.

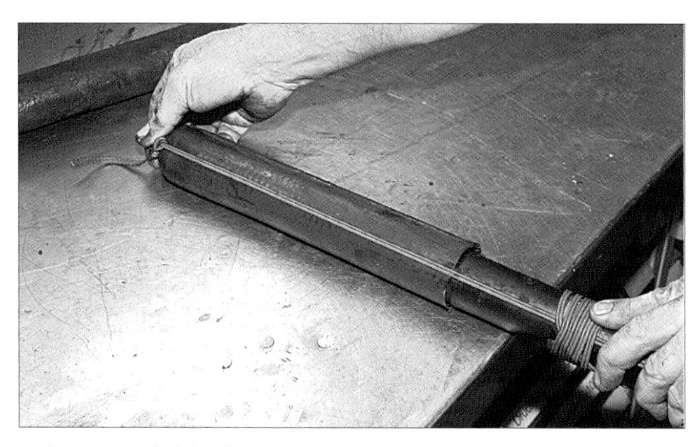

To fit the new unit, it is advisable to secure the shock absorber in its fully compressed state.

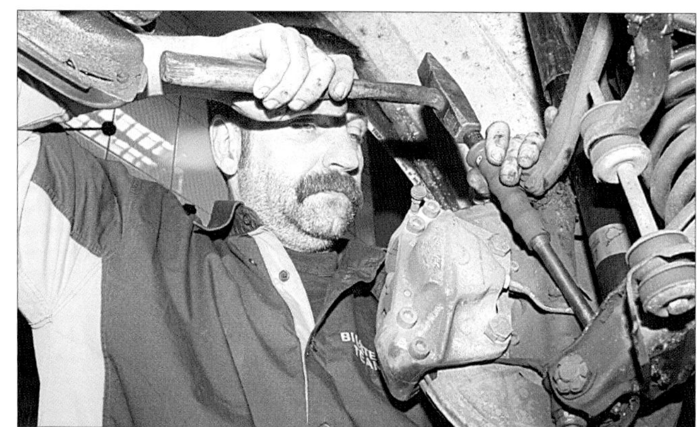

With a couple of well-aimed blows of the hammer, the expert makes sure that the lower mounting ...

... slips into position. Then he tightens the self-locking nuts.

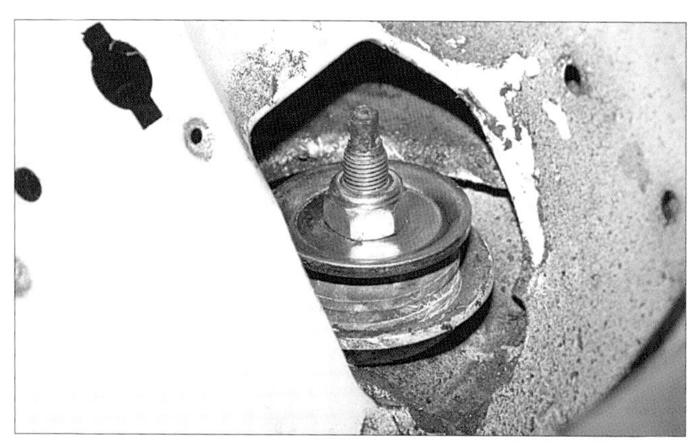

At the top too, instead of two nuts, one self-locking nut is employed.

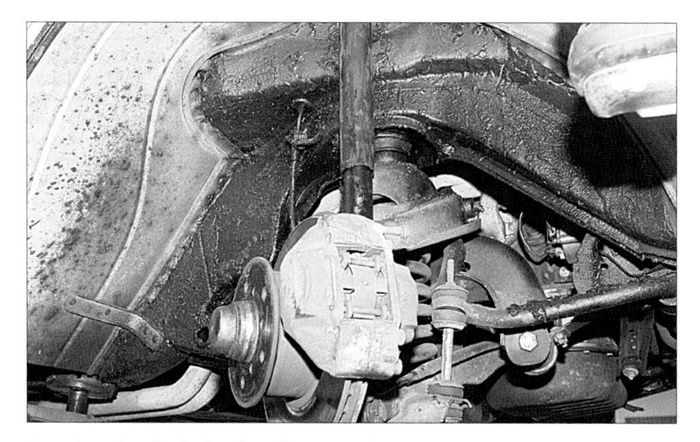

To replace the shock absorber, the expert didn't even need to remove the wheelarch liner which had been added to the car. Including removal and refitting, the job was done in half an hour.

shortened shock absorbers, for which the piston reverts to its correct central position after the suspension has been lowered.

MacPherson struts occupy an exceptional position in the realm of gas-filled shock absorbers, as one of the few disadvantages of the mono-tube design becomes apparent. Since the working tube of the shock absorber must be manufactured quite precisely, it is virtually impossible to weld any additional parts onto the outside of it. With a MacPherson strut, the spring mount and steering knuckle form a single unit with the shock absorber. Bilstein has made a virtue of necessity and developed its own model of MacPherson strut. In Bilstein's version, a mono-tube shock absorber with its piston rod at the bottom (ie upside-down) is placed in a guide tube, which once again functions as the carrier for the steering knuckle and the spring plate. The trick of this design is that the damper tube slides up and down inside the guide tube and so absorbs the lateral forces produced by the suspension unit. Unlike the standard design, none of the force used to control the wheel is applied to the piston rod, which ensures that the shock absorber responds consistently regardless of the load applied to it.

Improving suspension performance is not, however, the only reason why it may be necessary to replace a car's shock absorbers. High sensitivity to crosswinds, premature lock-up of the brakes, and uneven tyre wear all indicate that the shock absorbers are worn out. Taking a Mercedes W108 as a case study, the foreman at Bilstein's Service Shop in Ennepetal showed us what to look out for when replacing shock absorbers.

Finding replacement parts for the German cruiser is no problem; after all, Bilstein has supplied original parts to Mercedes since the 1950s.

After 250,000km (160,000 miles), the shock absorbers came through the one-off test during the German safety inspection (TÜV) once again with no significant faults found, but they had seen better days. In addition, when

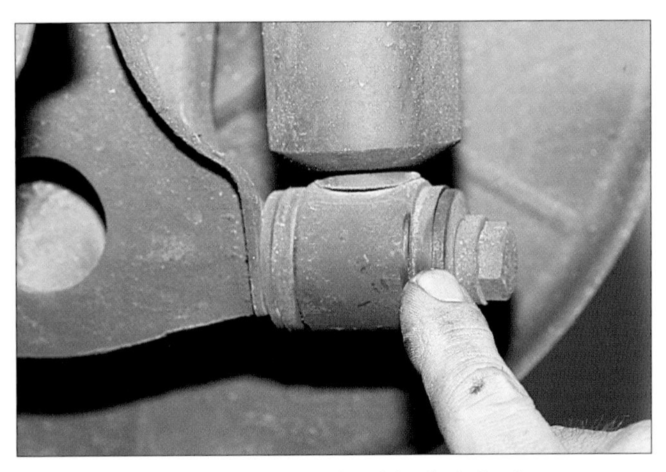

The loose bolt on the bottom mounting ring of the shock absorber was responsible for the ominous knocking sound from the rear axle.

The foreman took the opportunity to give all the suspension bushings a thorough check. Everything is in order here.

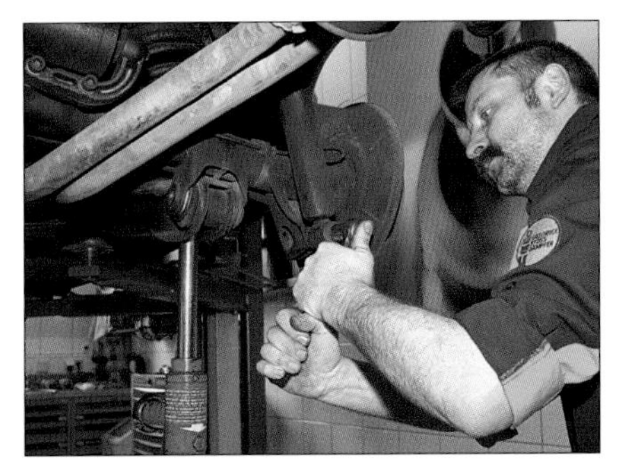

At the rear of the car too, the axle was supported with the gearbox lift, before ...

... the simple attachment to the lower shock absorber ring could be released.

In the boot, at the top of the shock absorber, however, the expert used his special tool to free it.

driving over manhole covers or other rough surfaces, a slight knock could be heard from the right-hand side at the rear of the car.

The fitter took the opportunity with the car up on a lift to examine all the rubber bushings used throughout the suspension, as quite different faults can often make it necessary to change the shock absorbers. In this case, as it happens, the lower shock absorber mounting ring was loose, a fault which could be put right at no cost with two turns of a 17mm wrench.

When replacing shock absorbers, it is helpful to have an hydraulic gearbox lift and a stepladder. But if you are fond of doing floor exercises in the gym, you can also do the job using a jack and axle stands. The expert used the gearbox lift to push the axle as high as possible from underneath, so that the springs on that axle were supporting part of the car's weight. Only then was the shock absorber relieved of its role in limiting the spring travel, and it could be unscrewed at top and bottom. From above, in the engine compartment (this is where the ladder comes in), the amateur mechanic should above all resist the temptation to release the shock absorber with a pneumatic screwdriver. It's not unheard of for the retaining screw of the working piston to come loose, sending the piston rod shooting through the workshop and causing serious injuries! To release the upper mounting, the specialist used a special socket to hold the piston rod in position while undoing the nut. This can also be done using an open-ended spanner and a box spanner, it just takes a bit longer. Then the shock absorber can be pulled out from underneath, assuming that you have enough space (which is where the lift is useful).

On other models of car, for reasons of space, the shock absorber is fitted in the middle of the coil spring. Sometimes it can be necessary to secure the spring with a spanner and to remove it together with the shock absorber. Before fitting the new shock absorbers, lay out the old and new units beside each other, so that you can make sure that you actually ordered – and received – the correct items. It is often easier to install them if you compress the shock absorbers and secure them in this position with a length of cable or cord.

The replacement job on the Mercedes' rear swing axle was uneventful, as the shock absorber was attached close to the spring. Before the new shock absorbers were tightened to the correct torque setting, however, the specialist had another important tip to share: "It is essential that the car is standing on its own wheels before you tighten the bolts at the bottom. Otherwise, the rubber bushings of the shock absorbers will be fixed in this position and will twist as soon as the springs on that axle are compressed. The rubber won't cope with that for long and will break away from the mounting ring after a few hundred miles. As a side-effect, the shock absorber piston rod will be operating under slight tension and the working piston will wear more quickly – a minor cause, but a major effect!"

There is even more room to work on the rear suspension of the old S-Class.

Watch out! With the car on the lift, the lower bolt should only be tightened by hand.

The rubber bump-stop, metal disc and nut can be fitted with ease here.

To tighten the self-locking nut, the fitter resorts to a conventional ring spanner.

Only when the car is back on its wheels should the bolt be fully tightened.

Even if the suspension is being lowered, the bottom mountings should be left undone and only tightened again once the suspension in in its new position.

A set of four standard-quality shock absorbers for the Mercedes cost about ●580. Sports shock absorbers are not available off the shelf for this sedate saloon. In principle, however, virtually any Bilstein shock absorber can be examined, overhauled and modified. An inspection, along with an hydraulic fluid change, starts at about ●60 for each shock absorber, the amount increasing in relation to the amount of work involved. All the way through to individually road-testing specific models, pretty much everything is possible in the customer special orders department. "We deliberately offer this service for owners of classic cars; ultimately, as the original parts supplier for many years, Bilstein has an obligation to keep these classics on the road."

160
HELLA
M ST 1292H

160
HELLA

Springs and shock absorbers

Diagnosing problems, replacement and improvements

Springs and shock absorbers do a hard job inside the wheelarches. If they develop a fault, the car's handling soon becomes dangerous. So, does stiff always mean good? Well, no thanks. Rather than giving them sporty handling, fitting lowered suspension has made many cars a pain to drive, as they ground when going over every kerbstone. Roadholding, safety and comfort are all dependent to a large extent on the springs and shock absorbers, and sometimes, too, your right to register the car as an unmodified, historic vehicle.

The set-up at the rear is simpler: the spring and shock absorber are usually separate.

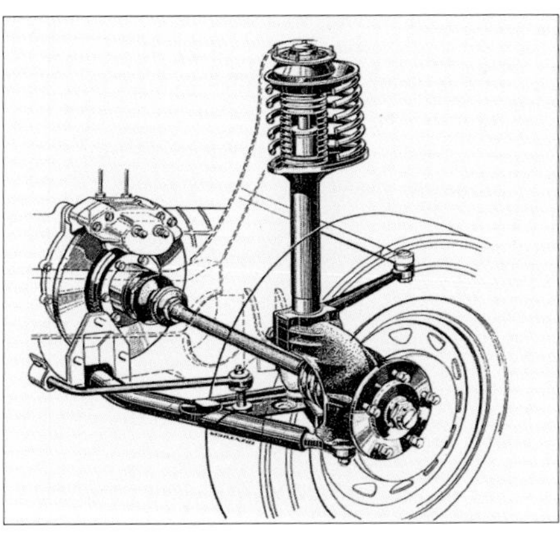

In the case of the MacPherson strut, the shock absorber and spring form a single unit, secured by its top mount.

The suspension is what connects the car's body to the road. As well as taking care of changes of direction when steering, its job is to reduce the oscillating motion of the wheel (which is part of the vehicle's unspring weight) and the bodywork in as controlled a fashion as possible. The transfer of these movements to the body should be avoided, in order to reduce roll and pitch, to minimise the dangerous build-up of vibration, and to ensure the best possible traction and grip, with as little slip as possible. With no springs, a vehicle would be undriveable on an uneven road. Working together with the shock absorbers, the springs ensure that the wheel remains in contact with the road. They absorb the forces produced when the car is driven, from mild vibrations and slight changes in inclination to the jolts and pitching motions which occur during sudden braking or steering inputs. In this situation, the shock absorber ensures that the spring, which, after reacting to a bump in the road, wants to return to its extended position with considerable force, is gently braked. This prevents the build-up of vibration in the car's body.

In principle, a shock absorber is a device for the conversion of energy, which transforms the kinetic energy inside the shock absorber into heat through a process of friction. Precisely calculated orifices in the shock absorber piston slow the piston rod as the hydraulic fluid passes through the shock absorber, so that all the vertical knocks and vibrations are reduced to a greater or lesser extent. The relative strength of the compression and rebound cycles is, on average, in a ratio of 25% to 75%. It follows that, after removing a unit which is in working order from the car, it is easier to push an hydraulic shock absorber together than to pull it apart.

With historic cars and modern classics, the most common types of shock absorbers are mono-tube and twin-tube models. The mono-tube shock absorber operates with the help of a gas charge. Although it is popularly referred to as a gas-filled shock absorber, in fact it contains hydraulic fluid. The tube filled with fluid inside the shock absorber is constantly maintained

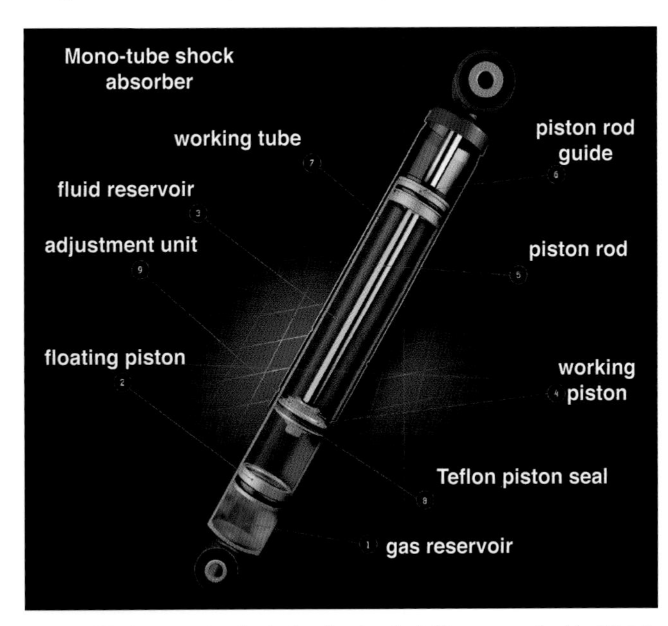

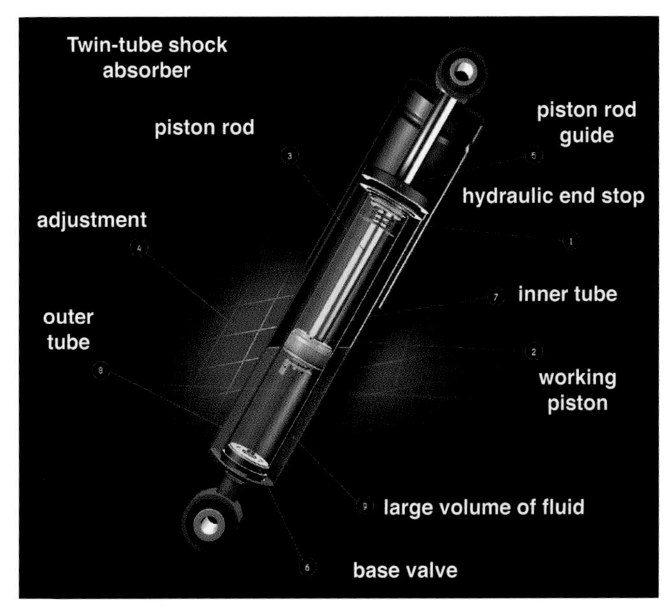

The gas-filled mono-tube shock absorber (on the left) was conceived by Bilstein and is popular today. The twin-tube shock absorber (to the right) has been considered state of the art since the 1950s and is generally referred to as an hydraulic shock absorber. It also exists with a gas charge in the outer tube, known as a twin-tube gas-charged shock absorber.

under pressure by an additional chamber containing nitrogen gas, which is separated from it by a floating piston. Bilstein was the first company to introduce gas-filled mono-tube shock absorbers.

Twin-tube shock absorbers only use hydraulic fluid. During the compression phase, the fluid is forced through a base valve from the inner to the outer tube; during the rebound phase, the valve again controls the flow of the fluid back into the inner tube.

These so-called telescopic shock absorbers, which are mounted separately from the spring, do not control the wheel location in any way.

Sports shock absorbers with suitably matched springs are often sold as a complete kit.

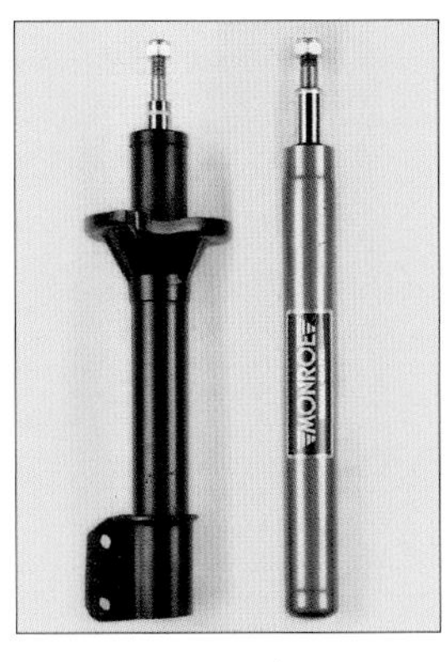

MacPherson struts have a cartridge insert. Some also form a single assembly with the shock absorber.

The MacPherson struts, which came to dominate car suspension design from the 1970s, provide a contrast to this approach: the shock absorbers (generally a cartridge insert), spring and upper wheel control are combined in a single unit with a top mount. While a telescopic shock absorber is only called on during the compression and rebound cycles, a suspension strut has more work to do: it contributes to the job of controlling the wheel location and is subjected to additional lateral forces during cornering, braking and acceleration. If the struts are replaced or overhauled, the camber and tracking should be checked afterwards.

When MacPherson struts are fitted, coil springs are always used. Coil springs have also been fitted in most other types of suspension since the 1950s. Until then, and in some cases until the present day, leaf springs and torsion bars were employed. These, likewise, rely on the addition of a shock absorber, but can also control the wheel location at the same time.

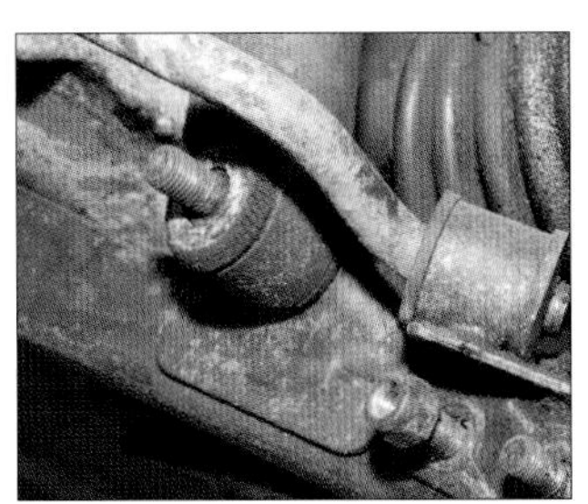

When old shock absorbers are removed, the typical damage often becomes apparent. The inner sleeve of the rubber mount rusts solid onto the shock absorber bolt, but can be removed using a grip wrench, as it will generally have split. Cracked or worn-out rubber mounts are often concealed by washers.

Whether you're dealing with struts or traditional shock absorbers, if the screw fitting is directly on top of the piston rod, you need to counter-hold the rod (left). The spring compressor is vital when changing shock absorbers and springs; without it, the job is impossible (centre). When replacing MacPherson struts, the top mount should also be changed (right).

Special spanners are available for the nuts on MacPherson struts. Sometimes, a pair of pliers will also do the trick.

Remember that the end of the spring must fit inside the moulded spring plate.

MacPherson struts are held under tension. The shock absorber units can also be screwed up tightly when taken off the car.

Dieter Glemser in a hurry, at the Nürburgring in 1973: the stiff rebound settings of his Bilstein racing shock absorbers, combined with hard springs and a thick anti-roll bar, account for the front wheel lifting off the ground.

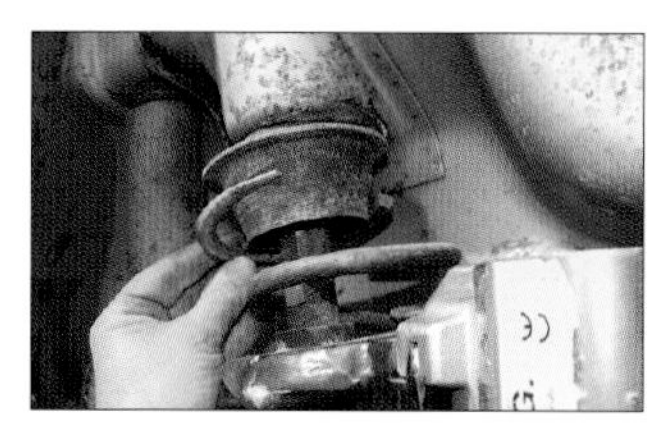

Cracked springs can occur as the result of metal fatigue or corrosion, and are usually found at the ends of the springs.

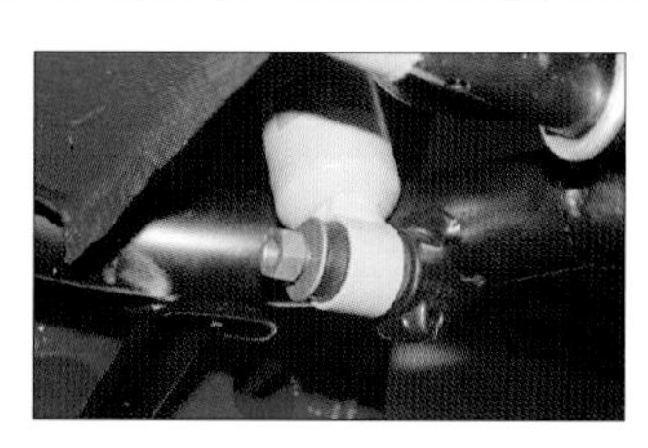

If you replace shock absorbers with the car on a lift, only fasten them tight after lowering the car.

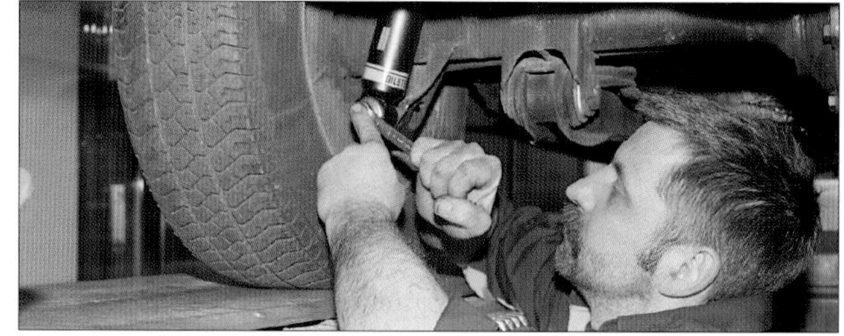

An uncomfortable job, but it's the only way: the complete set of bolts can only be tightened with the car's weight on its wheels. Otherwise, the rubber mounts will tighten and split within a short time, and you will have to start the work over again.

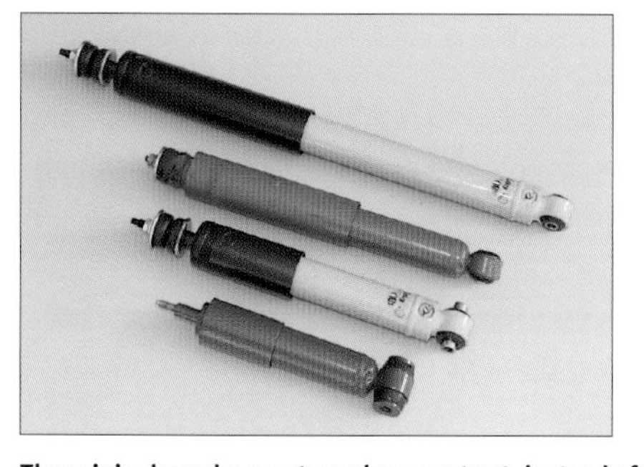

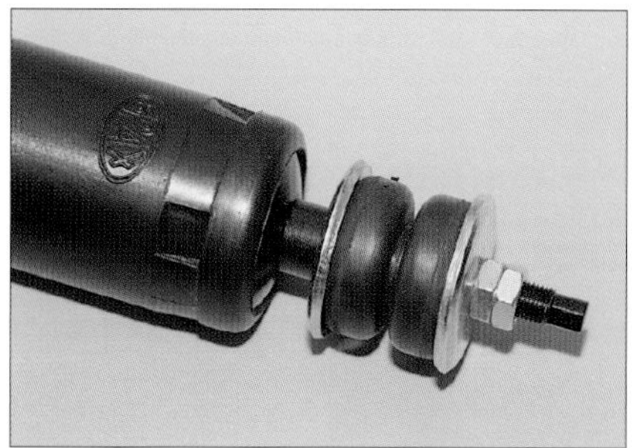

The originals and a sporty replacement set: instead of the green hydraulic shock absorbers, adjustable gas-filled shock absorbers can be fitted. Don't be put off by their modern appearance: what matters is how they perform. The rubber washers and possibly the end stops should also be checked or replaced.

Faulty springs can only be identified if visible cracks become apparent, most often at their ends. Severe corrosion is also a bad sign. Rust can develop very quickly (and with destructive effects) in the case of plastic-coated springs, such as those fitted to most modern classics. Until a few years ago, no primer was applied under the powder-based paint. If the protective coating comes away, rust immediately sets in. If a spring of this type subsequently breaks, as a rule it will be at one of these rusted areas. Leaf springs with slight corrosion are less of a problem: it is often enough to dismantle them, then clean, paint and grease them.

A broken shock absorber is easier to recognise. If fluid is leaking from the tube, then the shock absorber has had it. Worn-out or cracked rubber mounts and washers can also be detected during a visual examination. The notorious bounce test also gives a clear indication: if you press down hard on the wing of the car, the body should bounce back a single time. The bounce test is often a good way to identify faulty top mounts, too, as they can rattle when subjected to quick movements.

The best way to diagnose faults is, of course, a test drive, ideally to a test centre which can generate a performance chart for the shock absorbers. That need not cost the earth, and may also potentially expose worn-out springs, the damage to which may not be visible.

Come on down: the sports spring is a good few inches shorter than the original. A homologation certificate or reference number (see above) should be available.

If repairs are needed, should you also make a few modifications at the same time? After all, a car with lowered suspension can look pretty racy. In principle, of course, that is perfectly possible, but the job isn't a minor undertaking, particularly from a legal point of view. A

A borderline case: not all vehicle examiners will accept lowered suspension on a VW Beetle or Microbus – such conversions were not yet customary in the 1960s.

car's suspension generally represents the compromise between comfort and sportiness which the manufacturer has established, and on which the springing and damping will have a decisive influence. As a rule, lower, stiffer suspension means a loss of ride comfort; sometimes, the permitted load or even the number of seats (four instead of five) will have to be reduced, and a statutory inspection certificate may be required.

In principle, lowering the suspension always entails fitting shorter springs. With many classic sports cars from the 1960s, that is no problem, as conversions such as these were carried out in period. In the meantime, a new generation of cars has become eligible for registration as historic vehicles: cars for which sport suspension kits were already offered 30 years ago.

These kits are therefore historically correct and do not threaten the cars' classic status.

The situation is different in the case of tuning options which only became popular more recently. Fitting anti-roll bars to Porsches or a VW Beetle is a well-known change, but the adjustable lowered front suspension for air-cooled cars from Wolfsburg and similar dropped suspension kits for Volkswagen's Microbus are met with scepticism by many engineers. A positive decision is really at the discretion of the individual vehicle examiner. After carrying out a conversion like this, you should always be aware of the risk that you may lose the historic vehicle status which your car previously enjoyed, or never obtain it in the first place. So, before planning to lower your car's suspension using more recent parts, you should check with the relevant authorities in your city or country.

If you use shorter springs, be clear that the shock absorbers must be adapted to suit them. If the standard shock absorbers limit the rebound travel, in the worst case it is even possible, when taking a corner quickly, that the (excessively) short sports spring will jump right out! If in doubt, a spring travel limiter should be fitted. Ideally, the shock absorbers should also be shortened to match the length of the springs, as, even during the compression phase, the piston can strike the inside of the shock absorber and destroy it within a short period of time.

The ideal solution is, of course, to fit shock absorbers which can be adjusted for firmness, such as those which Spax or Koni have supplied for a long time. If you are only changing the shock absorbers without lowering the suspension, there is, of course, no need to go through the homologation process.

Only flying beats this! For many cars, sports suspension kits were already available in the 1970s: as period accessories, they are no obstacle to registering a car today as a historic vehicle.

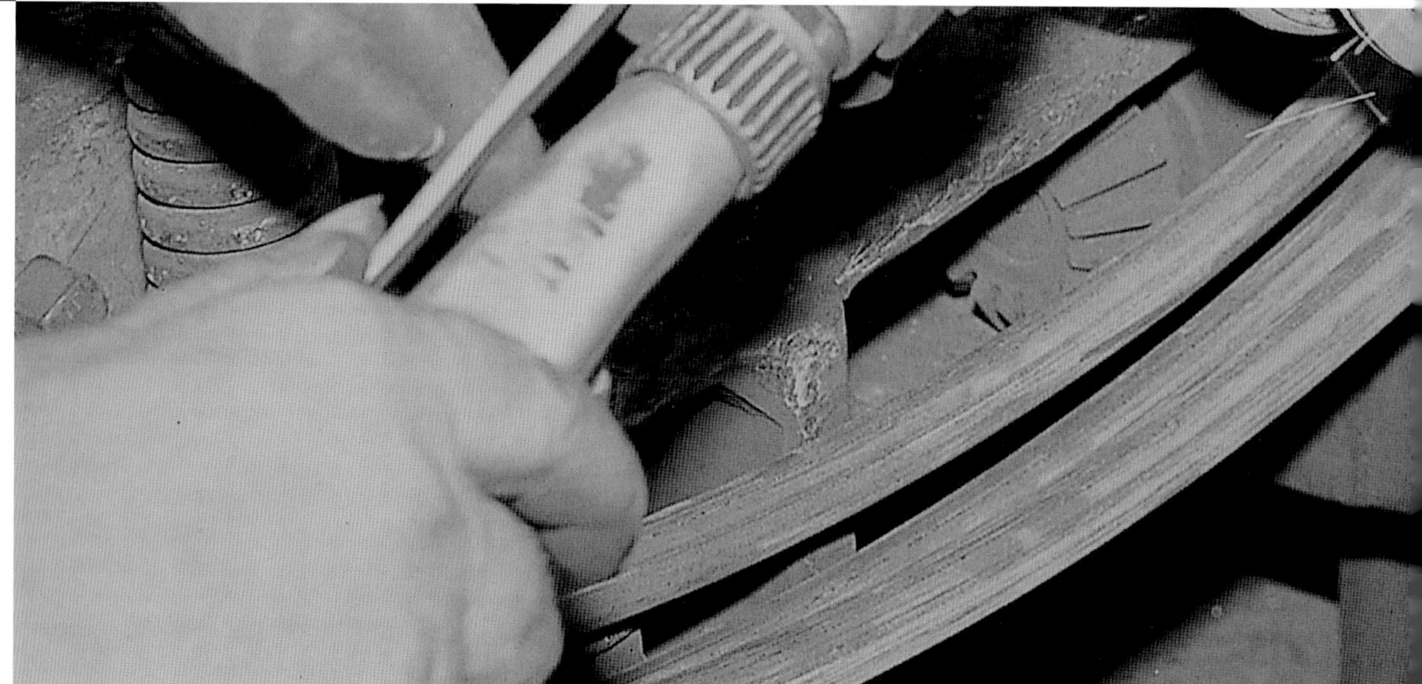

Springs
Types of spring and how they are made

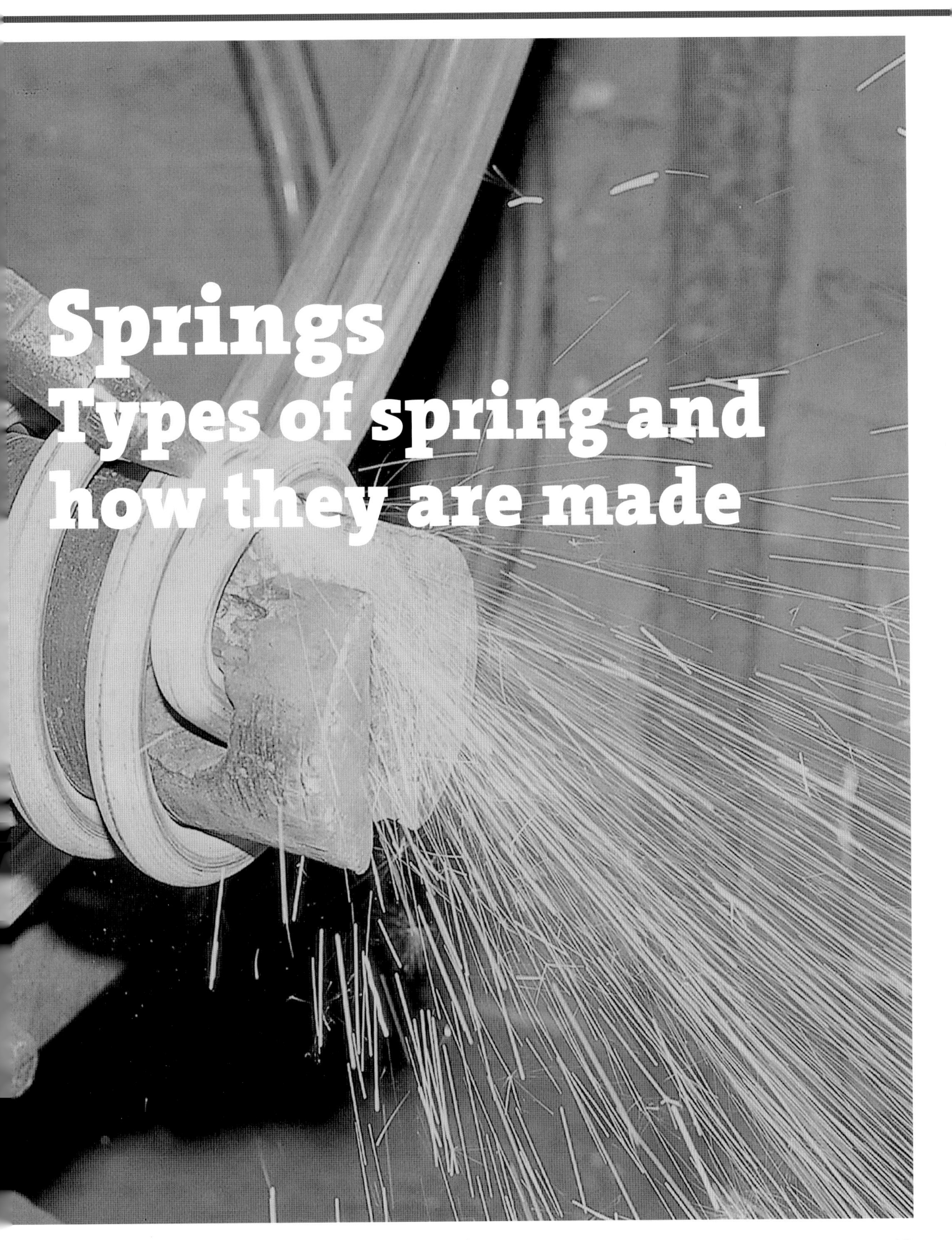

Springs are one of those chassis components to which little attention is normally paid. It is only when the time comes to replace them that much thought is given to making the right choice. It is even possible to produce one-off parts.

Simple though they may appear, these components have a considerable influence on a car's handling characteristics. With no suspension, the hard jolts which are felt when driving over potholes or bumps in the road would be transmitted directly to the chassis frame or body, placing an undesirable load on them. Suspension is not just a question of comfort. The connection which the springs, together with the shock absorbers, form between the car's axles and its body ensure that the wheels remain in contact with the road surface, while the body should, ideally, seem to glide over rough tracks.

The stiffness of the springs, the so-called spring rate, is defined as the relationship between the spring travel and the load imposed on them. The spring rate depends on the material used to produce the springs (we'll limit ourselves to steel springs in this instance), on its specific qualities (ie the quality of the steel) and on the exact design of the springs (their size and shape). In principle, all springs absorb the forces exerted on them as they become compressed. Naturally, the basic tenet of energy conservation also applies here: no spring can actually absorb the energy, which, in practice, can only be stored and then more or less quickly released, as the spring returns to its original form during the rebound phase.

In the pages that follow, we'll be concerned with the three basic designs of steel springs: coil springs, leaf springs and torsion-bar springs. All three types are nearly always made from a special spring steel, an alloy to which silicon and manganese have been added, and sometimes chrome and vanadium as well. Leaf springs are the oldest known design, but they have the worst energy absorption characteristics, so end up being relatively large and heavy. Torsion-bar springs are a very different matter: no other design of steel spring makes better use of the inherent qualities of the steel used than the torsion-bar spring, which can be thought of as a coil spring which has been unwound. Compact and easy to produce, coil springs represent a happy medium for car manufacturers, and have become the most widely-used design. We will therefore now turn our attention to these; subsequent chapters are devoted to torsion-bar and leaf springs.

Coil springs began their triumphant progress more than 100 years ago in the Swabian city of Reutlingen, with the development of the first winding machine which could be used for volume production. Since then, it has become impossible to imagine car manufacturing without them. Their great advantage is that they can absorb substantial loads while remaining light in weight and not taking up much space. Their secret lies in how the coils are wound: this brings the tensional strength of a long spring steel cable to a compact, generally cylindrical design. A significant advantage of coil springs is that – depending on how they are designed – they can be extended or compressed. They therefore present a huge range of possibilities to car manufacturers in solving technical problems. Springs are most often compressed; extension springs are typically used, for example, as return springs for carburettor control linkages or folding motorcycle stands. It is extremely rare, however, to find them used in car and motorbike suspension designs.

The thickness of the wire makes all the difference: the yellow spring on the left will support twice as much weight.

In mass production, coil springs first appeared after the Second World War, and, in the years that followed, they gradually displaced the leaf springs which had been customary until then. The combination of independent front suspension with coil springs and a live rear axle with leaf springs was also very popular. Nowadays, coil springs are considered leading-edge technology; it is only for commercial and off-road vehicles that leaf springs continue to play an important part.

Initially, coil springs were wound with a constant pitch (the spacing between the coils) and consistent diameter. In this case, the expert will describe the spring characteristics as linear: the force of the spring increases approximately in proportion to the spring travel. Springs such as these respond well on slightly uneven surfaces, but the body of the car has to move a great deal before the full spring pressure is produced.

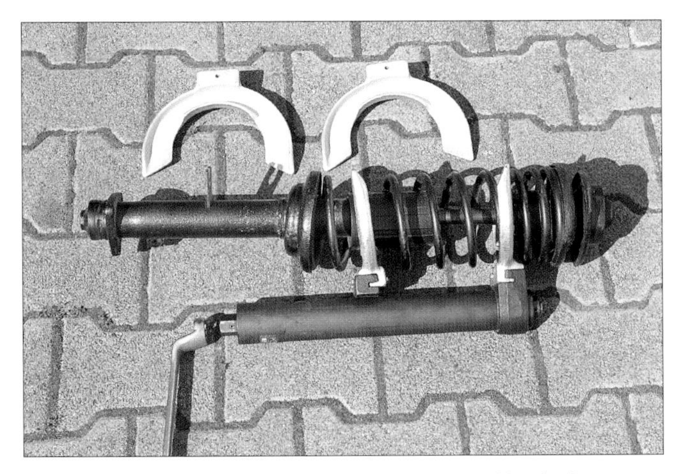

Pre-tensioning the springs corresponds to a force of several hundred kilograms. Removing them without the right specialist tools is extremely dangerous.

MacPherson struts cannot be disassembled without using a special clamping device. The clamp prevents the collar from slipping free.

Simple tensioning devices should have a transverse brace, which will prevent slippage.

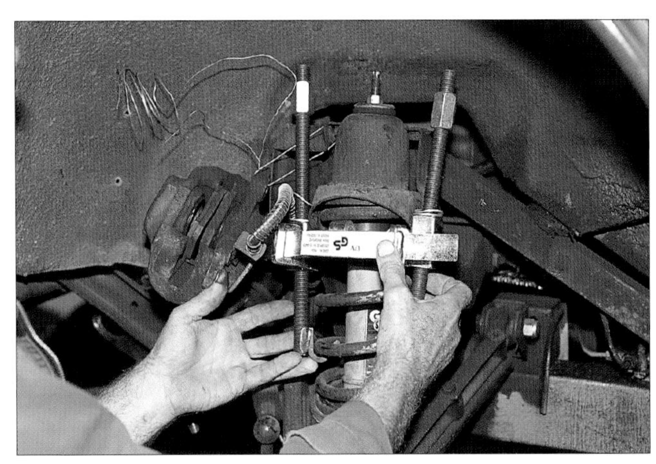

Universal tools cannot be used on all cars. In particular, space can be a problem on the front suspension of modern classics.

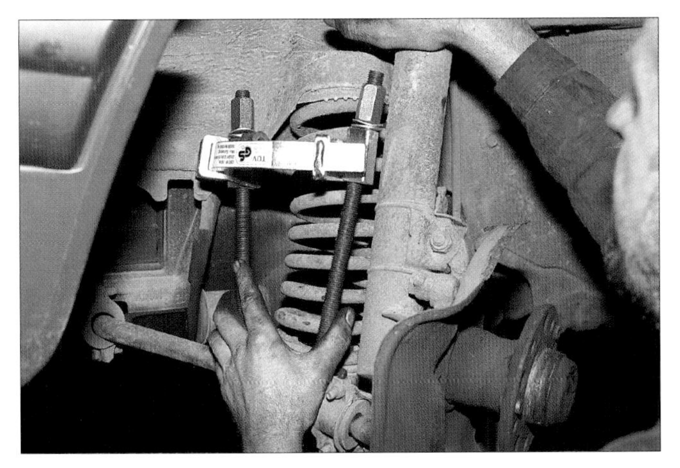

Here, too, the MacPherson strut gets in the way and the universal tensioner cannot be used. Only a special tool will do the job ...

... and this comprises two discs and a threaded rod. The rod is pushed through the lower disc and screwed into the thread of the upper disc.

The long suspension travel which results may be considered desirable for off-road vehicles, but is not at all suitable for car chassis intended for sporting use on the road. In order to achieve a sensitive response, together with an acceptable spring travel, the manufacturers hit on the idea of producing springs with a variable pitch. These so-called 'progressive-rate springs' provide a finely-tuned response over that part of their length which is tightly coiled, but their resistance quickly grows as the spring travel increases, since the section of the spring with a greater pitch then comes into play. The spring response follows an upward curve, starting gently and becoming

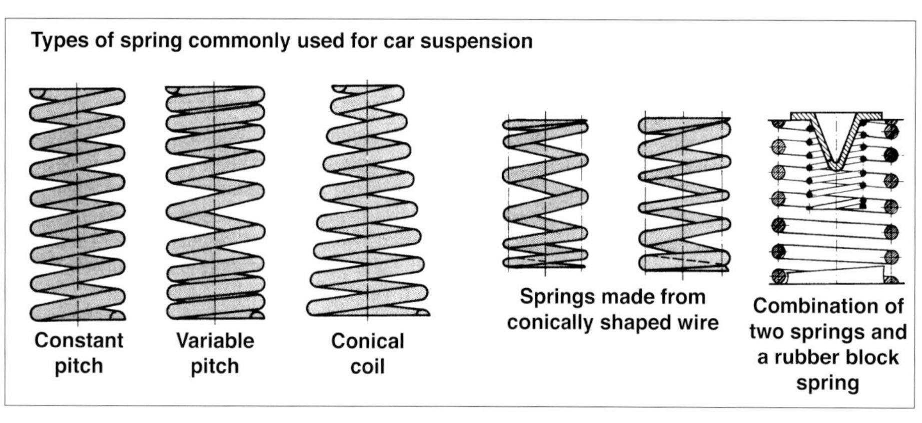

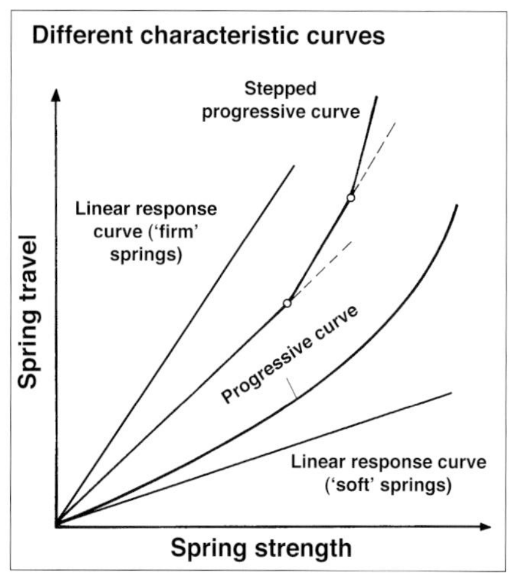

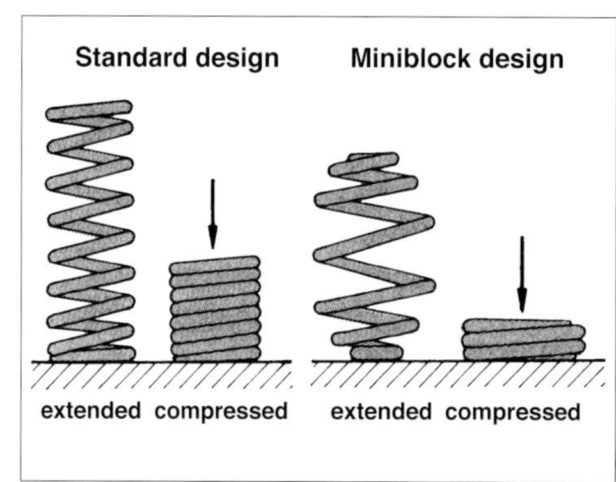

progressively firmer. The transition is seamless. In the meantime, progressive-rate springs have become an essential fitment on sports cars with short suspension travel.

Another approach to progressive-rate springing consists of using a conical spring with a constant pitch. Seen from the side, the cylindrical shape resembles a cone.

The force or rate of the spring is dependent not only on the thickness or gauge of the wire, the pitch of the spring and the characteristics of the steel used, but also the diameter of the coil. In the case of a conical spring, the large-diameter sections initially provide a gentle response to minor bumps in the road. When these have responded as far as they are able, the smaller-diameter sections take over, producing correspondingly more resistance. It is often difficult to distinguish the characteristic curve of a progressive-rate spring from that of a conical spring. Today, the leading manufacturers have obtained similar results from cylindrically-shaped springs with a constant pitch, but which use a different gauge of wire towards the centre.

A virtually progressive characteristic curve can also be achieved by combining two or even three springs. In our example (see the chart at the top of the page), the large primary spring reacts when the car goes over smaller bumps. Only when this spring has been compressed about halfway through its travel does the smaller secondary spring come into play, and the characteristic curve becomes markedly steeper. The rubber block springs function as the third and final element when the primary and secondary springs are almost completely compressed. The characteristic curve steepens once again. Since the graph shows three sections with the line at different angles, this is referred to as a stepped progressive curve.

Another common design of progressive-rate springs today is the so-called 'miniblock spring.' In this design, the pitch and the diameter of the coils both vary. As the diagram shows, their principal advantage consists of their significantly lower height when compared with a cylindrical coil spring. Whatever type of spring is fitted to your classic, all coil springs suffer from one minor disadvantage: in contrast to leaf springs, they produce virtually no intrinsic damping effect. In practice, the shock absorbers for an axle fitted with coil springs must always be somewhat larger than for an axle with leaf springs.

So much for the theory. Most mechanics only get to grips with the springs on their car when they need to be replaced. The most obvious reason for this is when the springs simply break. That occurs more frequently with some models of car than others, although the exact cause for this cannot always be clearly determined. Alfa Romeo's Giulia range is notorious in this respect, as its rear coil springs fracture quite often. Experienced Alfa owners make do by fitting the strengthened springs designed for cars used for towing or an additional central spring available from Italian aftermarket suppliers. Insiders consider the problem to be caused by the standard shock absorbers, which have a much smaller compression travel compared with the rebound. Large bumps in the road literally put such a strain on the springs that they hit their rubber bump stops. In truth, this problem arises far less frequently with cars fitted with modern, gas-filled shock absorbers.

Another reason for changing the springs on classic cars is in cases of metal fatigue. Cars such as these often manage to keep going when they are lightly loaded or are subjected to only moderate shocks from the road. Occasionally, the problem can be detected from the car's sagging bodywork. In general, workshops do not have exact data against which they can check car springs, let alone the appropriate measuring devices. In marked contrast to valve springs, for example, it is extremely rare to find information in repair manuals about the length of springs in their uncompressed state (ie when they have

The spring steel wire is delivered in the form of large reels.

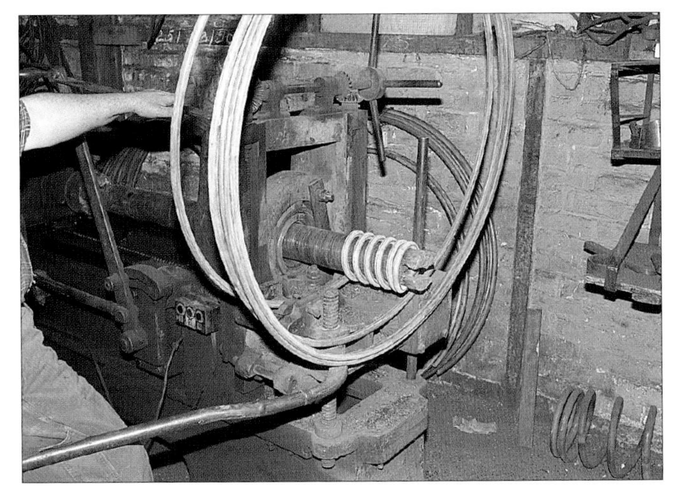

By adjusting the feed, progressive-rate springs can be wound.

Different diameters of spring can be produced, depending on the size of the core.

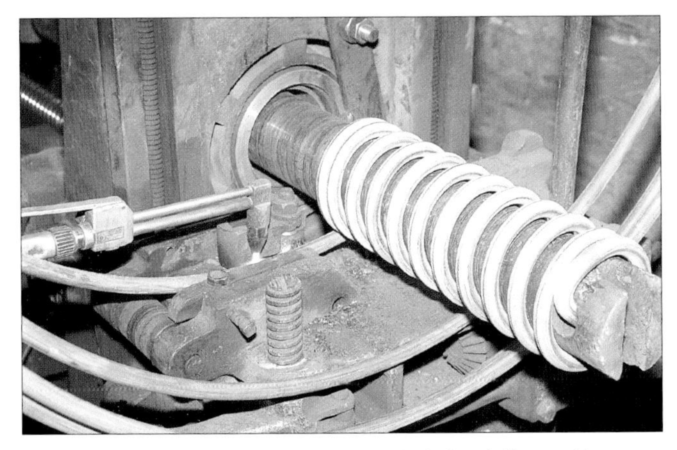

As soon as the prescribed length has been reached, the winding machine stops automatically.

To harden the spring, it is first placed in a furnace heated to 850°C (1560°F) for about 20 minutes.

been removed from the car). If the measurements do, however, reveal that the spring is shorter than it should be, then clearly it will have to be replaced.

If you notice that the springs on your car are not performing well – if they are too soft or often bottom out, for example – you should seriously consider replacing them. The damage which could ensue – such as cracks in load-bearing parts or accidents caused by poor handling – could be much more expensive, and, above all, extremely dangerous. The obvious thought when replacing the parts is to take the opportunity to improve your car's handling. That is not, however, so straightforward, as springs and shock absorbers were originally developed with great care to suit each model of car and how it would be driven.

Lowering the suspension in particular can often be detrimental to the original set-up, if the shock absorbers and their mountings are no longer working at the optimal angle.

In the light of current traffic conditions, there is no reason why the original suspension set-up should not be changed to suit your personal preferences. The change above all from linear to progressive-rate springs often significantly improves a car's handling. Firmer springs are available from specialists for many classics. They should present no problems during mandatory safety inspections, such as the MoT test in the UK, as they are normally supplied with official certification. Incidentally, it should also be mentioned that undesirable body roll in corners will scarcely be reduced by fitting different springs: a stronger anti-roll bar would be the best choice here.

When changing the springs on your car, however, you need to keep your feet firmly on the ground, as the load applied to the car's bodyshell increases unnecessarily when fitting stiffer springs.

In any event, low-slung cars with very hard suspension are not really any faster on twisting country roads. If the springs and shock absorbers do not allow the wheels enough movement, they lose their grip on the road. A car which has been set up with softer suspension could carry a higher speed through corners. The moral of the story: in most cases, there is no need to be mercilessly shaken about in your seat. But at the end of the day, suspension changes remain a matter of individual taste, as long as they meet your country's legal requirements.

One piece of advice: cars sold in the Middle East often had firmer springs because of the poorer roads there, and these parts can sometimes still be found with the help of a friendly storeman. In many cases, though, you will have to make do with fitting the standard springs to your classic. Thisis by no means the worst solution, since who really drives their classic on its door handles today, unless they are taking part in an historic race? For cars used in competition, there is a wide choice of springs available.

But what can you do if no springs whatsoever remain available for your car? That was the question we asked ourselves, as we looked for a specialist who would not baulk at making one-off parts. It was not such a simple matter, as the best-known manufacturers understandably prefer to produce at least a small batch of springs. For single-make car clubs, who might require 100 or 200 springs, that makes perfect sense. Eventually, we found

As long as the spring is glowing red hot, adjustments to the length and pitch can still be made.

Next, the spring is quenched in an oil bath, which makes it brittle and hard.

Above all, the length must be exactly right, or the car will sag to one side.

So that the spring will later sit flat inside the mounting cap, the ends are ground smooth.

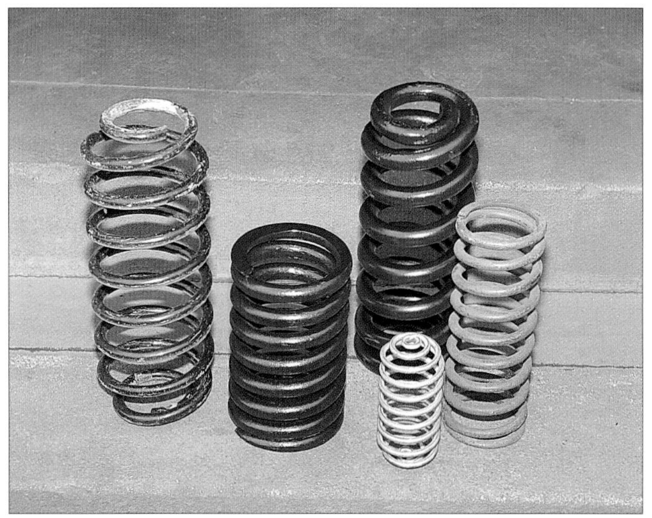

Even one-off jobs are possible: the specialists can fabricate almost any spring, with a minimum gauge of 5mm (0.2in). Springs fitted on the same axle should always be replaced as a pair.

a firm in Dortmund, whose owner looked after a few old motorbikes of his own, and was very open to working with the classic car community. He deals with all types of spring, and, on this occasion, we were able to look over his shoulder as he and his employees produced some coil springs.

The raw material used is either a pre-hardened spring steel wire (67SiCr5-type alloy) or an unhardened wire (50QV4), which will only be hardened after the spring has been wound. The strength of wire required for car springs comes from large continuous reels with a diameter of at least one metre. For a gauge of up to 25mm (one inch), the cold metal is wound on the machine directly around a suitable metal core. The feed rate, which can be constant or variable, is configured in advance on the machine. Wire with a gauge greater than 25mm (one inch) is heated until it is red hot before it is wound. Once the machine has completed the appropriate number of coils around the core, the wire is cut off using a welding torch.

Many springs are levelled off at the top and bottom, so that a flat spring cap can be used. In Dortmund, this job is carried out by hand, using a large grinding machine. Depending on the metal used, the springs are then placed in a furnace at a temperature of 850°C (1560°F) for about 20 minutes.

While they are red hot, the length of the springs can still be corrected. Next, they are quenched in an oil bath. At this stage, they are so brittle that they would immediately snap if subjected to a load. To make the springs pliant again and to give them their definitive characteristics, they are once again placed in a furnace, where they are heated to a temperature of 450°C (840°F) for an hour. The expert refers to this process as 'tempering.' Depending on the customer's wishes, the springs can finally be painted or coated in plastic.

For this specialised company in Dortmund completely unknown springs are no problem, even if they have fractured. The spring rate can be determined from a broken section of a spring, and the pitch can be measured. If the right material is chosen from the start, winding the coils is a quick affair. Coil springs should, however, always be replaced as a pair on the same axle, so spring specialists always fabricate at least two springs. Customer requirements such as 'Make them 10% firmer' are no problem when manufacturing new units. If the customer wishes, the metal used and the quality of the springs can be certified, so that, as a rule, their inspection by the German TÜV testing organisation poses no problems.

We were surprised by the price of the newly manufactured springs: with a delivery time of at least three weeks, springs for cars cost about •x65. For larger quantities, the prices come down. As well as leaf and coil springs, the specialist in Dortmund can make pretty much anything which can be bent into shape to work as a spring. While we were there, a package arrived containing 30 or so extremely varied types of spring for Lloyd Alexander cars, which a club wanted to reproduce. "Those will need a good deal of work by hand," the expert calmly observed.

MS VW356H

MS VW356H

Springs
Lowering the suspension

The realisation that a lower centre of gravity has a positive influence on a car's handling is almost as old as the car itself. It was only natural, therefore, that, at some stage, individual owners should begin to lower their cars' suspension. For a time, it was especially popular to adjust the spring plates on the VW Beetle in order to bring the torsion bars closer to the ground.

You can't argue about taste – or so the ancient Romans thought. When the then rulers of the Mediterranean coined this phrase, however, no cars existed. And what better subject is there to argue about than a car's appearance?

Let's face it: sitting there now, after the spring plates (known as 'swords' by German VW owners) have been adjusted, the Volkswagen takes some getting used to. Its tail squats low on the ground, and the rear wheels have so much negative camber that it almost looks as if the slightest load would make the whole lot collapse at any moment. The Beetle's nose, meanwhile, stands up at a jaunty angle. Although the road is completely level, you might well think the car is going slightly uphill.

VW drivers are confronted with a well-known problem: since the Beetle was never designed to cope with high lateral forces, the oil pressure warning lamp regularly comes on when cornering quickly. A further disadvantage of lowering the suspension on Beetles is that the car is more prone to hit its rubber bump stops.

Of course, you don't have to opt for such a radical course of action. If you are not put off by adjusting the inner splined coupling (which requires removing the wing and using a protractor to measure the change), you will be able to choose from a wide range of settings. If you wish, you can, of course, also raise the suspension on the Beetle. One sensible option is to rotate the inner coupling by three splines anticlockwise (on the right-hand side of the car), and the spring plate on the outer sprocket by three splines clockwise. Since there are 40 splines on the inside and 44 on the outside, this corresponds to lowering the suspension by 1.8cm (0.7in). In this case, the car's load capacity remains unchanged. With sportier set-ups, where the suspension is lowered by 4cm (1.6in)

It seems hard to believe, but the Beetle was once a popular sports car for amateur racers. Without lowering the rear end, the race was already over.

The swing axle of the VW Beetle: the torsion bars with their inner and outer splines can be clearly seen.

However unusual the visual impression given by the car may seem, it used to be common to lower the rear suspension of a VW Beetle or Karmann Ghia. Lowering the centre of gravity and changing the camber settings counteracted these rear-engined cars' tendency to oversteer. Ferdinand Porsche's swing axle design, with torsion-bar springs and lengthwise spring plates, made it easy to lower the suspension and to reverse any changes made.

The simplest way to lower the rear of the Beetle is to rotate the spring plate by one spline (clockwise on the right-hand side of the car, anticlockwise on the left). The process illustrated in the sequence of photographs in this chapter takes little more than an hour for each side of the car, and results in a drop of about 8cm (3in), which is pretty neat. It makes an enormous difference to the handling: any tendency to oversteer in normal driving disappears, and corners can be taken appreciably faster. Then as now, or more, the car will only be allowed to carry four rather than five passengers. Whenever lowering the suspension, it is important to remember to only tighten the shock absorbers by hand when the car is jacked up, and to tighten them fully once the car is on the ground and the wheels are supporting its weight.

Naturally, it is also possible to lower the Beetle's suspension at the front, but this is a much bigger undertaking, as the assembly has to be modified using a milling machine and welding apparatus. In addition, there is little tradition for doing so, and certainly not in motorsport. As late as the 1980s, according to Gerd Hack and Dieter Korp in their cult handbook, *Now I'll make it go faster*, there was no question of lowering the front suspension. So-called 'lowered axle assemblies' had in fact existed in the United States in the 1960s, but only came to Germany when customising the cars' appearance became more and more popular.

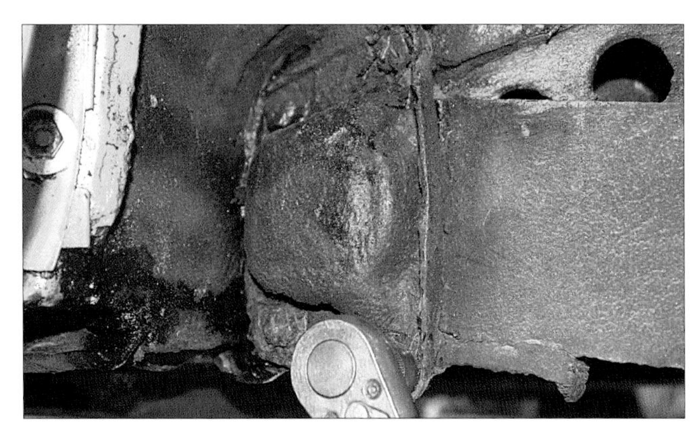

First, the cover over the spring plate mount is removed. As you do so ...

... check the condition of the rubber bushes and, if necessary, replace them.

It is important to mark the initial setting. 'L' indicates the left-hand torsion bar.

The bearing flange for the axle tube and shock absorbers has to be undone.

Removing the spring plate may require some effort.

There are 44 splines on the outside of the torsion bar and 40 on the inside.

Hardened steel: in principle, the torsion-bar spring is a coil spring which has been unwound.

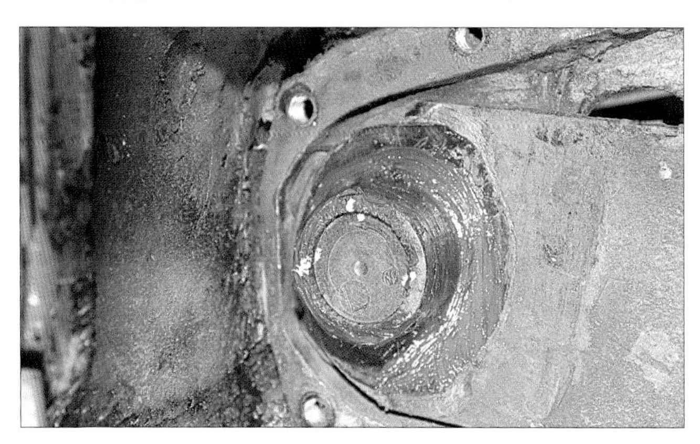

The spring plate is rotated by one spline anticlockwise (on the left-hand side of the car).

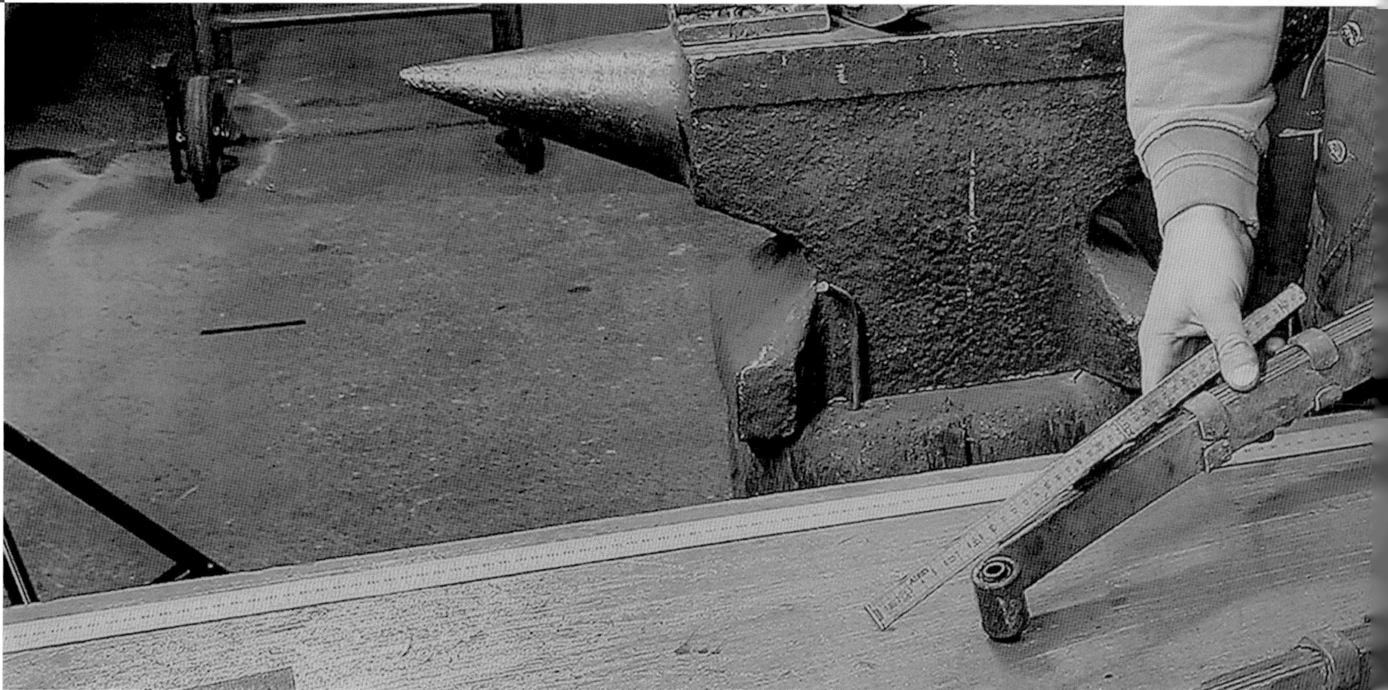

Springs
Repairing and replacing leaf springs

Straightening 'tired' leaf springs; now there's something every classic car enthusiast has heard about at some time or other. We asked an expert how the job should be done properly, and, also, how you can influence the firmness and shape of the springs, in order, for example, to discreetly lower the suspension of a classic car.

In car building, leaf springs typically consist of several 'leaves,' whose number, length and cross-section determine the suspension characteristics. The greater the number of leaves, the firmer the suspension will be, and the greater the load it can bear. An extremely soft suspension can be obtained with just three or four leaves. A normal leaf spring displays a linear characteristic curve. A progressive response rate, however, in which the suspension becomes disproportionately firm when subjected to an additional load, can only be obtained by combining several different sets of springs.

In comparison to other steel springs, leaf springs have to contend with some significant disadvantages: on the one hand, their high weight and cumbersome dimensions; and on the other, the intrinsic damping caused by the friction between the leaves. This makes it harder to achieve a sensitive response from the suspension. In contrast to coil springs, however, leaf springs can take over or contribute to the job of locating the axle.

At this point, we would like to explain how 'tired' leaf springs can be given a new lease of life, and how their firmness and shape can be modified, in order, for example, to unobtrusively lower the suspension of a classic car. A little Austin A30 – which was already listing slightly to one side – was an ideal patient for our ministrations. A suspension overhaul was urgently called for, after which the owner of this little British car wanted to pep up its engine.

Armed with both leaf springs, we went to visit a specialist in Stassfurt, near Magdeburg. This company produces small and medium-sized batches of leaf springs for all manner of applications, but is happy to deal with individual requests from drivers of classic cars. Even looking at them for the first time, the expert found problems which had not been noticed until then. With no load on them, the leaf springs were of different lengths (what the specialist refers to as their 'span') and also of different height (specialists use the term 'camber' to describe this measurement of the arc of the spring when it is at rest). In addition, the experts in Stassfurt checked the behaviour of the springs on a test bench. With the help of an hydraulic press, this determined how much travel the springs showed in response to a given weight. The result: for both springs to compress by 1cm (0.4in), a load of 30kg (66lb) was required. With a compression of 3cm (1.2in), however, there was a difference of 15kg (33lb) between the two values (55-70kg/121-154lb), and with a compression of 10cm (4in), a difference of 20kg/44lb (160-180kg/353-397lb). The stiffness of the springs at the rear of the Austin, therefore, varied over part of their travel by as much as 25%!

The next step depends on the condition of the individual leaves once they have been taken apart. Dismantling them is straightforward: the leaf spring is held securely in a vice and the spring clips are freed. On some types of spring, rivets may have to be removed, but with others, it's enough to bend open or unscrew the clips. Next, the centre bolt, which holds the leaves of the spring together, is undone.

The sobering result of the first inspection: the camber of one of the Austin's springs measured 113mm (4.4in).

The other, however, came in at 120mm (4.7in). The expert suspected that the springs originally came from two different cars.

This assumption was also supported by the difference in length of 7mm (0.3in) between them.

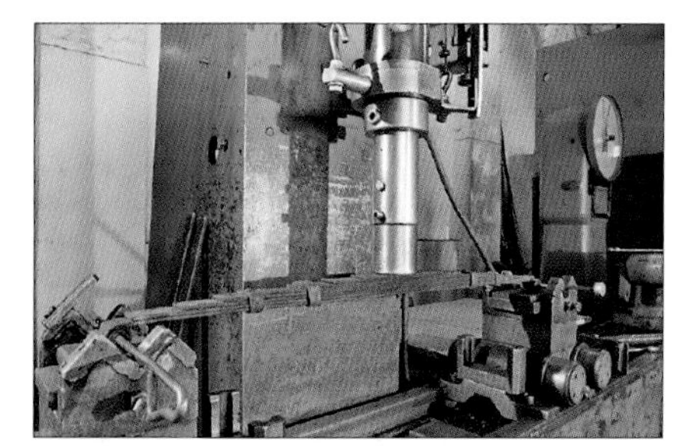

After the initial examination, and before being dismantled, the spring assemblies were taken to the test bench. An hydraulic press was used to exert force on the springs ...

... and the load in kilograms required to compress the springs was read off on a scale marked in one-centimetre increments.

In our case, the two springs produced different results, with the measuring device giving values of 160 and 180kg (353/397lb) for a spring compression of 10cm (4in).

The next stage in the work was to take apart the springs: on the Austin, two clamps were held in place with rivets.

The two clamps could easily be bent open using a large pair of pliers, which had been extended with a lever.

Before the centre bolt – which holds together the set of leaves in the middle – is undone, the spring should be held in a vice.

When the clamps and centre bolt have been undone, the leaves will practically fall apart on the work surface.

Don't forget at this point to mark up the springs, especially if you are dismantling both of them at the same time.

The first clean-up at the end of the preparatory work: the wire brush is used to remove road dirt and loose rust.

After this, the individual leaves can be cleaned and examined on the workbench. Where the ends of the shorter leaves rest on the longer leaves, they should not have created an indentation, and, above all, the 'load-bearing' leaf with the spring eyes should not have lost more than 10% of its original thickness.

Our springs were still in the green zone: in the section subjected to the weakest load, the load-bearing leaves were 4mm (0.16in) thick, and next to the spring eyes, where the next shortest leaves are situated, 3.6mm (0.14in) thick. Two further tests are reserved for professionals: the first is an electromagnetic crack test using a special fluid and UV light, while the second consists of measuring the stiffness of the springs with a test device.

The manager of the firm in Stassfurt was categorical: "As long as there are no cracks or mechanical damage, virtually any leaf spring can be restored to its original shape and stiffness. We refer to this as straightening or opening up the spring. This is nothing more mysterious than shaping the individual leaves with a hammer when they are cold. It sounds easy, but demands a good deal of experience, as the shape of the different leaves must, of course, fit exactly against each other, and because usually no documentation is available for the original springs. In the end, therefore, just how much the springs need straightening comes down to experience." If the springs need to be straightened to a very considerable degree, it is essential to 'temper' them afterwards in the furnace. This consists of heating them evenly to a temperature of about 500°C (930°F), and then letting them slowly cool.

In the case of the Austin, however, the intention was not to return the spring to its original shape, but to lower

This check is reserved for professionals: an electromagnetic test using a special fluid ...

... makes even the smallest cracks visible in UV light. There are no ifs and buts about it: cracked leaf springs should be scrapped.

A matter of experience: a few well-aimed blows of the hammer give the load-bearing leaf ...

... its desired form, which is then marked up in chalk on the work surface.

Not quite there yet: no gap should be visible between the leaves.

When shaping each leaf, it is important to work from the middle outwards.

Nice work: when the individual leaves sit flush against each other, as shown here, the job is done.

Here the camber is also being checked to make sure the measurement is correct.

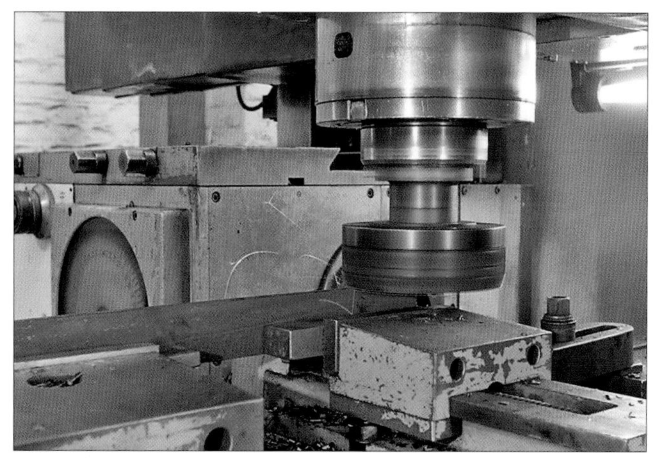
Producing a new leaf spring: the 4mm-thick blank is cut to length and milled down to 38mm (1.5in).

Giving the new leaf its final shape requires plenty of heat. It is therefore heated until it is red hot in a furnace at about 1000°C (1830°F).

When the glowing leaf emerges from the furnace, it is roughly cast into shape ...

... and then quenched in a special oil bath. Only the gradual process of tempering it ...

... at about 500°C (930°F) in a second furnace will make the spring pliant again.

the car by 3-4cm (1.2-1.4in). The process is the same, whether the springs need to be straightened or forced back together. The expert turned to a 3kg (6lb) hammer with a rounded head. In general, the load-bearing or 'guide' leaf is always the first leaf to be formed on the anvil. The shape of all the leaves which succeed it will follow this one.

It's important always to start from the middle of the

To prevent the test results from being distorted due to friction between them, the leaves are greased.

It's always best to check: examining the set of leaves as they are provisionally reassembled.

A light blow with a hammer helps straighten out individual leaves.

Under pressure: the second testing device shows how much firmer the spring has become.

Feeling the strain: this is how the spring looks when it is compressed by 10cm (4in).

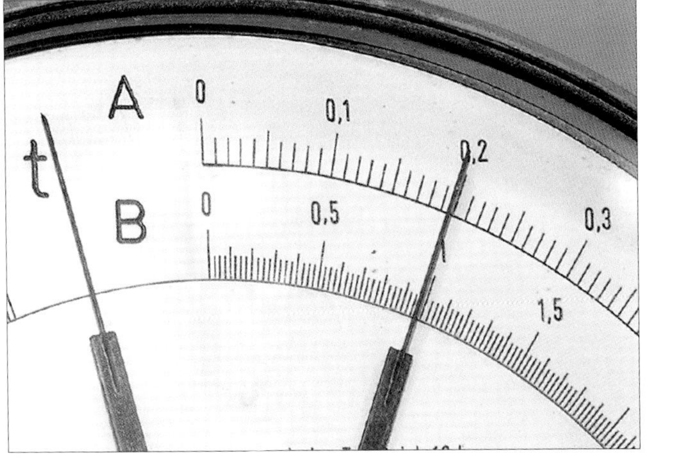

With an additional leaf fitted, a load of 200kg (440lb) was needed.

spring and work evenly outwards, regularly checking your progress. After half a dozen or so well-aimed blows, the specialist set the load-bearing leaf to one side and drew its contour in chalk on a metal work surface. That makes it easy to check that both sides of the spring are consistently shaped when they are turned through 180 degrees. If everything matches up, the camber of the load-bearing leaf is then measured to see if it is close

Making notes as you go: the behaviour of the spring throughout its travel was recorded in ...

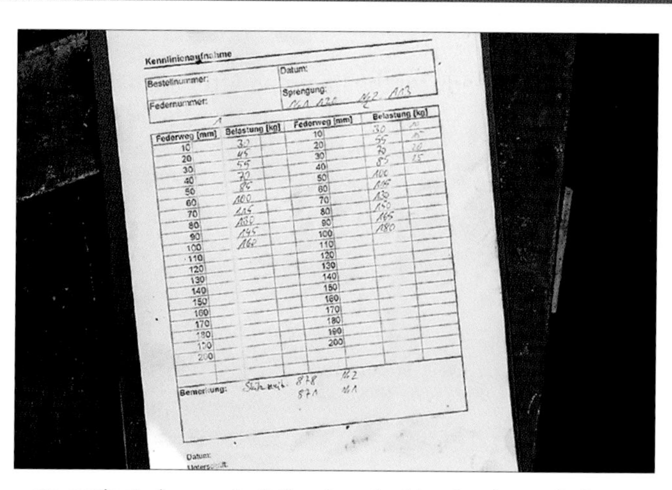

Kennlinienaufnahme

Bestellnummer:

Datum:

Federnummer:

Sprengung:

Federweg [mm]	Belastung [kg]	Federweg [mm]	Belastung [kg]
10	30	10	30
20	45	20	55
30	55	30	70
40	70	40	85
50	85	50	100
60	100	60	115
70	115	70	130
80	130	80	150
90	145	90	165
100	160	100	180
110		110	
120		120	
130		130	
140		140	
150		150	
160		160	
170		170	
180		180	
190		190	
200		200	

Bemerkung:

Datum:

... one-centimetre increments. Both springs should perform in exactly the same way.

Firmer, and lower too: the new camber measurement is 85mm (3.3in), 35mm (1.4in) less than before.

The last stage in the job lies ahead: if everything fits, the leaves will be sandblasted, painted and then finally fitted back on the car.

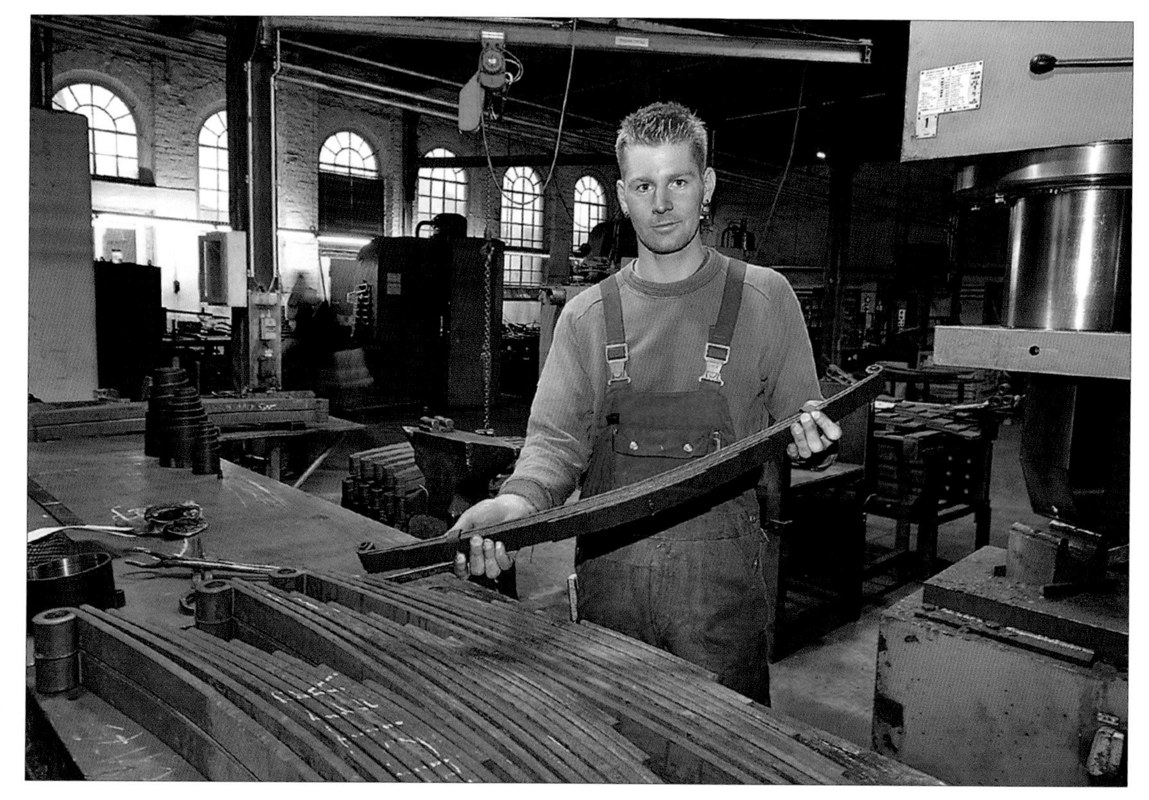

Ready for fitting: with a fresh set of springs, the little British car should handle crisply.

to the target. If everything so far is in order, the next shortest leaf is ready for the hammer. After each couple of blows, the expert compared the shape of the leaves. It is important that the shorter leaves should always be rather more steeply arced than the longer ones, and this should be kept in mind when wielding the hammer. To be certain that the leaves will sit perfectly on top of each other when under load, the expert compared them by clamping them together.

Three or four goes was usually all it took before he was satisfied and moved on to the next leaf. When the final leaf had at last been shaped, the entire set was clamped together, and the camber measured once again. In this case, only half of the 4cm (1.6in) reduction in the camber that we wanted had been achieved, meaning that all the leaves had to go under the hammer a second time ...

We also wanted the springs on the Austin to be a bit firmer, which was not a problem, as the proprietor explained to us: "Usually, we'll do this by fitting an additional leaf, which should be as long as possible. The dimensions of this extra leaf are determined in relation to the existing leaves. On the Austin, the leaves are 38mm (1.5in) wide and 4mm (0.16in) thick.

"We have some 4mm (0.16in) blanks to hand, which are 40mm (1.6in) wide, and have set up a milling machine so that it is easy to remove 2mm from them. In principle, this is how we produce any leaf spring, even if only a single leaf in the set is broken. Usually, the most difficult thing is when we receive vague requests from the customer. 'Lots harder' isn't much for us to go on. In general, we reckon that 10 or 15% stiffer springs already have a noticeable effect."

But even when the leaves had been cut to length and milled off, the job was not yet done. The new leaf must be given its final shape – and that required plenty of heat. The length of spring steel was placed in a blazing furnace at just over 1000°C (1830°F), and then bent roughly into the same shape as the other leaves it would sit alongside. Next, while still red hot, it was quenched in an oil bath (for some spring steels a water bath is used). When it emerged, it was as hard as glass and brittle; it would break as soon as it was subjected to any load. At the end of the process therefore, it was once more tempered to some 500°C (930°F), making the spring pliant again. After it had cooled, the hammer was used again to ensure a perfect result.

Before the leaf springs were finally sandblasted and painted, they were checked once more on the test bench. In our case, this confirmed exactly the results we were looking for: on both sides of the car, a compression of 10cm (4in) required a load of 200kg (440lb), and the camber was 85mm (3.3in) rather than 113/120mm (4.4/4.7in).

For all the work illustrated here, as well as changing the bushes in the spring eyes and shortening the longer load-bearing leaves (which required re-making both spring eyes), the company we used for our case study charges about •x180 for each leaf spring – a very fair price, given all the work involved. The test drive should prove how worthwhile it had all been, and we were already looking forward to it ...

Axle assemblies
Overhaul and repair

If your car has failed a vibration test, like that in the German TÜV safety inspection, then there's work to be done. The vibration test rig is merciless in showing up the smallest defect in a car's steering or suspension. It can help identify faults which often go undetected by an amateur mechanic. Even the early stages of wear can be uncovered, which is all to the good, as there are few parts of the car where the individual components need to work together so closely as in its suspension. Damage to one component will quickly trigger another fault elsewhere. If the ball joints have had it, the increase in the force applied when steering will mean that the tie rod ends will not last much longer either. Misaligned tie rods will, in turn, result in uneven tyre wear. If the kingpins are faulty, the whole wheel will start to wobble.

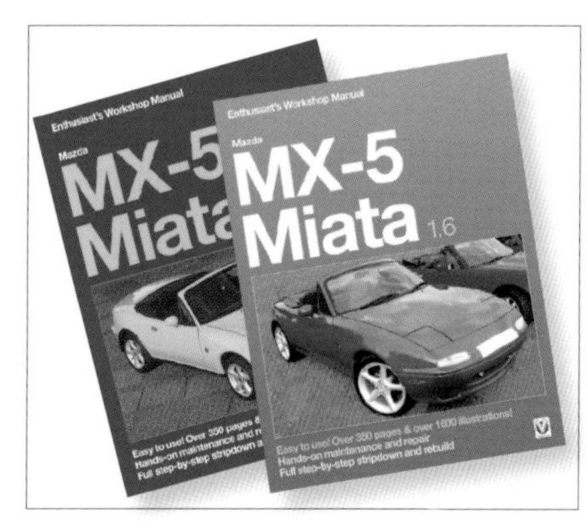

Without a repair manual you won't get far, as you always need the alignment settings.

The inspection keeps your car safe, prevents more problems in the future, and ultimately contributes to your pleasure behind the wheel!

The vibration tests essentially check three sets of components: the ball joints; the suspension bushes; and, on older models, the kingpins. We'll explain the job these different components have to perform, and what can go wrong with them, before embarking on our workshop tour.

The earliest designs of car suspension used a solid front axle. The wheel, complete with the steering knuckle, was held in place with a bolt with a brass bearing. After

A typical example: the TR6 has two wishbones on each side.

Dismantling starts with the anti-roll bar attachment.

A puller is used to pry off the tie rod ends.

In the past, so-called 'pickle forks' were used, but they weren't great so close to the rubber gaiters.

The lower retaining bolt can be withdrawn safely.

The shock absorber is still retaining the spring. Next, the steering knuckle can be folded out of the way.

The upper wishbone is not held under tension.

The steering knuckle can now be removed, together with the wishbone.

Further disassembly can be carried out comfortably on the workbench.

Take care: the load must now be taken off the shock absorber, so that ...

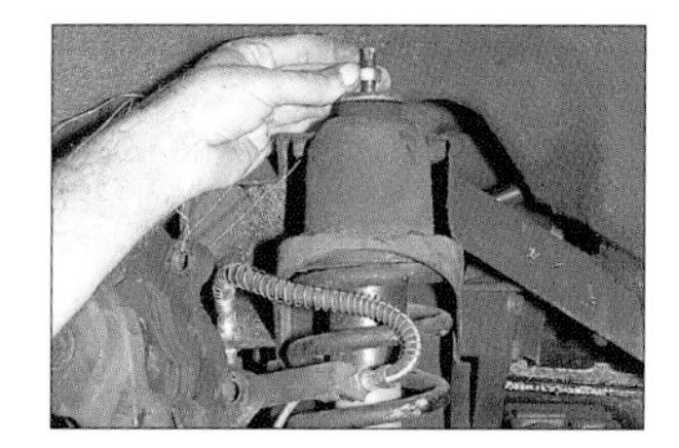

... the upper mounting bolt can be undone without danger.

Once the bottom bracket has been dismantled, the shock absorber ...

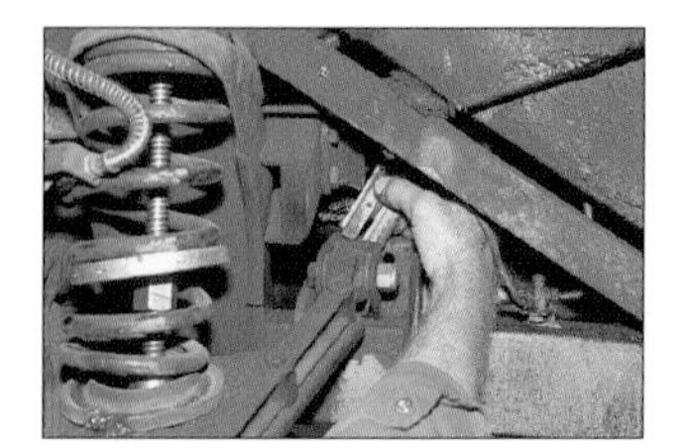

... can be withdrawn. Now the bottom wishbone, together with ...

... the camber adjustment plates, can be removed with the help of a spring compressor.

On the workbench: the lower wishbone link needs to come out.

Then the bracket on the upper wishbone can be undone.

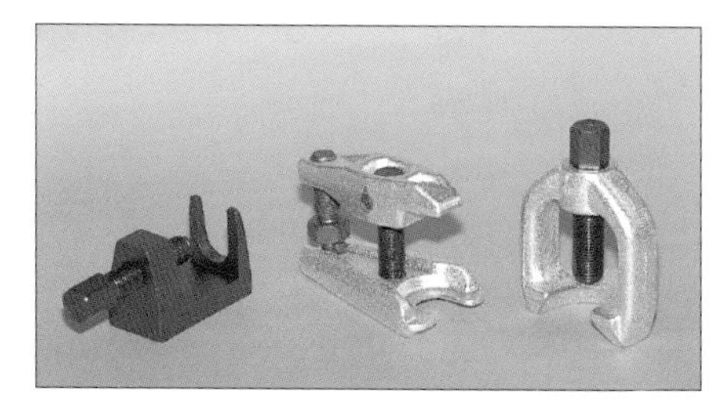

You won't get far working on car suspension without the tools to pull or press out parts.

Above: the different front axle components.
Right: expensive, but good-quality. This complete set of tools contains everything you need to work on the suspension.

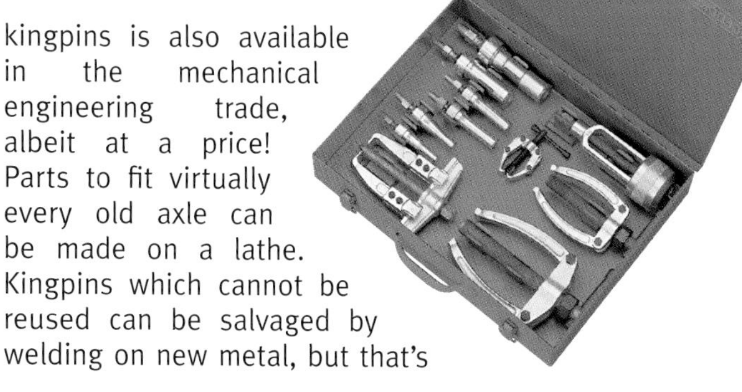

the Second World War, independent suspension became increasingly well established. Over the course of time, the kingpins were replaced by maintenance-free ball joints, which provided the lower and, depending on the design, upper mounting point for the wheel. In passenger car design, the upper ball joint mounting has virtually disappeared nowadays, and the MacPherson strut has long been predominant.

Until the 1960s, however, the most common method of suspension used kingpins. These functioned in much the same way as a hinge: a vertical pin was rigidly connected to the axle mounting, and the steering knuckle swivelled about it. In order to reduce the effects of friction, the pin was located in a brass bushing. For it to work smoothly, it had to be kept constantly greased, or the bushings would, sooner or later, wear right down. Properly maintained, mountings using kingpins can last forever, although the brass parts, will usually not last as long as the car.

Unfortunately, these days the replacement parts needed to carry out repairs are gradually becoming harder to find. Contacts in single-make car clubs can be helpful, but sometimes the only solution is to have parts specially made. It's not difficult to get hold of the bronze used to make bushings, while the type of steel suitable for kingpins is also available in the mechanical engineering trade, albeit at a price! Parts to fit virtually every old axle can be made on a lathe. Kingpins which cannot be reused can be salvaged by welding on new metal, but that's

Two bolts hold the upper supporting link in place. Once they have been undone ...

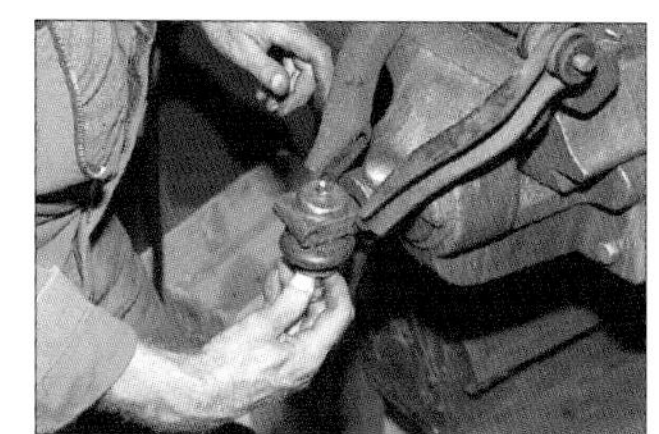

... the two halves of the wishbone can be removed from each side of the bracket.

During the suspension overhaul, the wheel bearings and brake discs can also be checked.

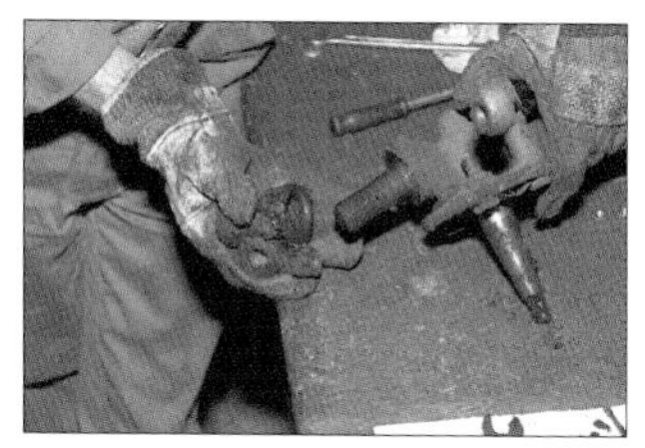

At the bottom, the steering knuckle moves in a brass joint that is resistant to wear.

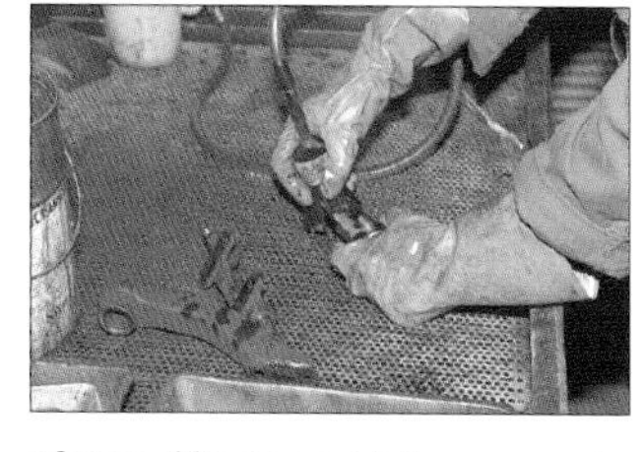

After soaking in a cold cleaner, most axle components look like new.

Take particular care over the position of any spacers and sealing rings used.

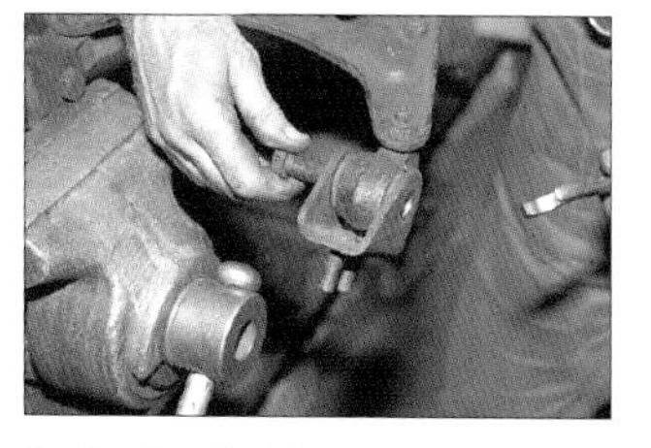

Gradually, all of the suspension components are completely dismantled.

It is best to push out the old rubber bushes with an hydraulic press.

Some bushes need a special press-fit adapter. The markings for such an adapter can be seen on these Volvo parts.

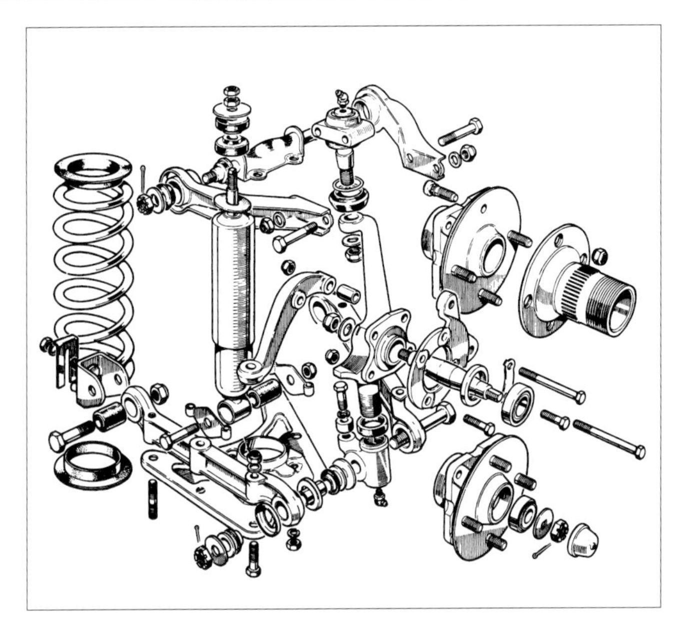

The comprehensive repair kit for the bottom wishbone on the TR6.

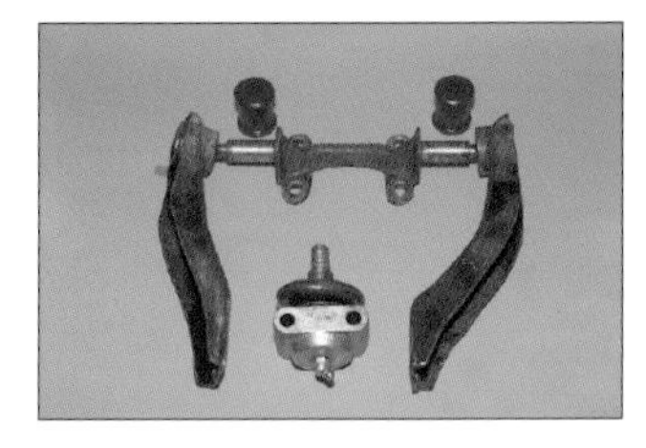

The layout of the top wishbone can be seen much more clearly.

Trying the new wishbone bushes for the first time.

The inner metal sleeves are pressed in separately.

Once the bushes have been pressed in, the new supporting link can follow.

Meanwhile, the bottom wishbone has also been fitted with new bushes.

Washers and sealing rings must be fitted in the correct sequence.

The brass joint is attached with a little oil on the stud.

a job best left to specialists. After this, the pin must be ground to size and the appropriate bushings fitted.

After they have been pressed in, the bearing bushings must be smoothed exactly to size using a reamer. The pin should not display any free play, but should turn smoothly. The hand reamers required are very expensive precision tools, and it is hard for a home mechanic to justify the expense. Instead, a good relationship with a metalworking expert can literally be worth its weight in gold. If you can bring them the parts with the bushings already fitted, they should be willing to finish them to size at minimal cost. For bushings that you have made yourself, you will need to add the lubrication grooves and grease pockets later by hand, so that the grease can make its way from the grease nipples to the actual mounting points.

The ball joint brought the era of the kingpin to an end. For motor vehicles, its introduction was almost as important as the discovery of the wheel! On 27 November 1922, the Düsseldorf company Fritz Faudi GmbH registered the patent for the steering ball joint. The maintenance-free ball joint, which was developed from it and is still used today, was introduced in 1952. It consists of a tapered bearing stud, which sits inside a round casing. Inside this, there are plastic shells, in which the spherical end of the pivot is located. In the production process, the pivot is inserted through the back of the casing and protected by means of a pressed cover. The diameter of the casing is smaller where the stud exits it, so that it cannot slip out in the event of damage. The ball bearings are permanently lubricated using molybdenum grease, and protected by means of a rubber gaiter. The ball joint developed by Faudi did not originally have plastic shells or a sleeve, but had to be greased at regular intervals.

For use with steering components, such as tie rods, maintenance-free ball joints quickly caught on across the world from 1952. During the course of the 1960s, ball joints gradually replaced the kingpins used in car suspension. As they also have to support the weight of the car, links such as these tend to be larger, but are essentially designed in the same way as steering tie rod links. Damage to the ball joint always causes excessive play in the stud; it usually arises from bearing shells which have ground down or cracked, the loss of lubricant, or dirt which has made its way in through leaks in the rubber gaiters. The gaiter is the only part which can be replaced. The joint should certainly have no free play, and the grease should be regularly replenished.

Even when kingpins were still in use, independent front suspension became the standard design. Since the comfort of the passengers could not be allowed to suffer because of the firmer suspension set-up, the manufacturers chose to fit rubber bushes as connecting elements in the suspension, which provided a damping effect. They not only reduced the jolts transmitted from the road and movements of the car's body, but the degree of flexibility which this approach to locating the wheels provided may also have reduced wear to the suspension itself.

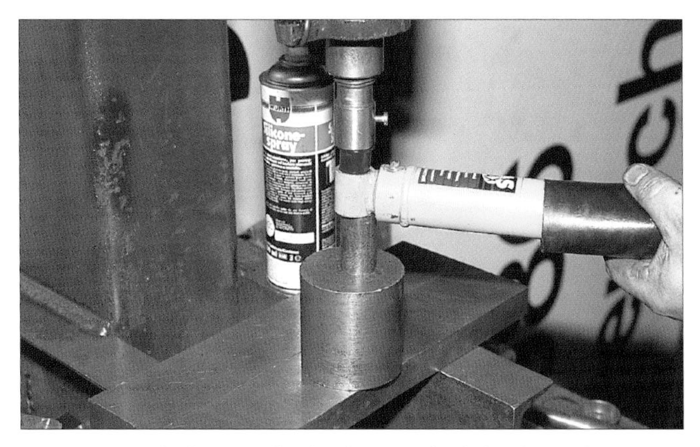

In our case, new bushes were fitted to the TR6's shock absorbers using a press.

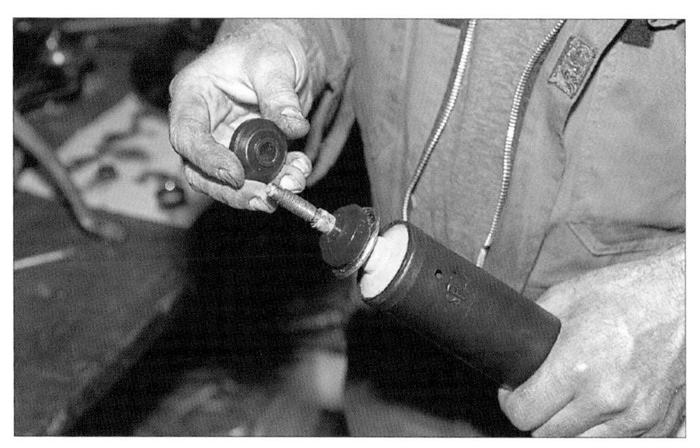

This professional also fits polyurethane bushes to the shock absorbers.

These so-called 'silent blocks' are among the hardest-working components in a car, and are continuously exposed to a high load, whatever the operating conditions. When the car is driven, they are subjected to constantly changing forces; when it is at rest, they bear the weight of the car. They have to withstand all this without any form of greasing or maintenance.

Silent blocks comprise an inner – and usually also an outer – metal sleeve, with a vulcanised rubber component to absorb vibrations. Silent blocks cannot be repaired. The rubber ages naturally and becomes either soft or brittle. The tough demands placed on them by heavy cars increase wear. In the event of damage, the blocks can no longer adequately absorb the forces they are subjected to, but transmit them in an uncontrolled manner, so allowing the different components to move excessively. Diagnosing the damage is not always easy, as the force which can be applied by hand or with a tyre lever to the axle of a car at rest is often insufficient to reveal a faulty silent block. If you're in any doubt, they should be replaced.

For many cars, including some classics, bushes made from polyurethane are now available, and will be fitted in one of our examples. This material ages more slowly, but is

The spring must first be compressed, before the bottom wishbone can be installed.

Only when the shock absorber has been completely bolted in place can the spring compressor be removed.

The shock absorber then bears the full pressure of the spring.

On this occasion, the tie rod ends were also replaced. Don't forget to get the alignment checked!

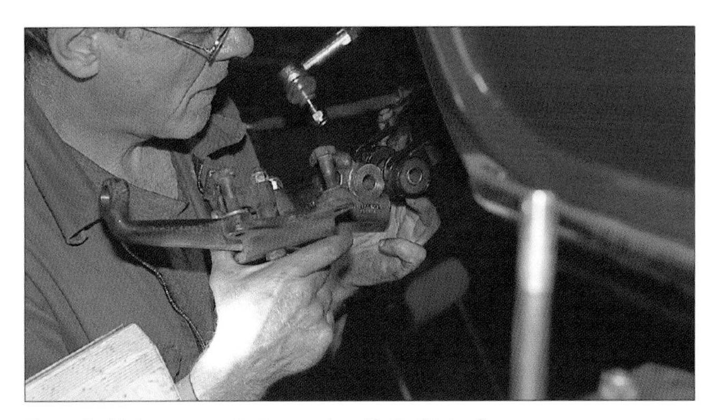

The swivel joint can now be inserted and bolted into place.

Next, the bolt on the upper supporting link can be attached to the steering swivel.

No problems when refitting the brakes after they have been checked and cleaned.

The final step is to tighten up the anti-roll bar.

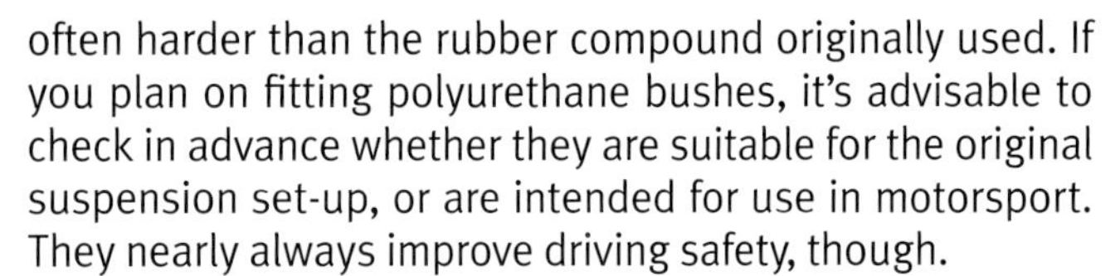

often harder than the rubber compound originally used. If you plan on fitting polyurethane bushes, it's advisable to check in advance whether they are suitable for the original suspension set-up, or are intended for use in motorsport. They nearly always improve driving safety, though.

Before we finally go into the workshop, we have one more important tip: overhauling an axle can often turn into a very tedious operation, particularly if you don't have the right specialist tools. It is, therefore, especially important to prepare for the job with the help of a repair manual, and to check the extent of the work and the tools required. If you embark on the job armed with plenty of patience and the right tools, everything should go well. Only the final – and essential – alignment check is a job for professionals!

To demonstrate overhauling an axle, we'll be using a Triumph TR6, a Mercedes W108 and a Volvo 245. We will also look at the front supporting strut on a VW Beetle. The work shown is broadly similar on most other postwar cars. The Triumph TR6, which we will turn to first in the workshop, is a typical example of a car with conventional independent suspension, using double wishbones. At the top, the steering knuckle is located in a ball socket, while, at the bottom, there is a short, threaded bolt in a brass swivel joint, sometimes also referred to in parts catalogues as a pivot bearing. The traditional kingpin is, therefore, no longer found here. The coil spring stands between the wishbones, attached at its base to the wishbone and at its top to the chassis frame next to the spring mount. The shock absorber operates inside the spring. Before dismantling the front axle assembly, take care that the shock absorber restricts the rebound travel and keeps the coil spring tensioned. Don't under-estimate the importance of this! It only makes sense to do this job if you have a suitable spring compressor. Pay no attention to suggestions to compress the spring with welding wire or screw clamps, which are a dangerous bodge: these flimsy affairs can easily burst open under the force of the springs. If your head happens to be inside the wheelarch at the time, that could prove fatal!

The photographs show the individual stages in dismantling the axle assembly. It is essential to support the lower wishbone from underneath, so that the shock absorber can be removed without being compressed. Once this has been done, the spring compressor (illustrated) can be placed in the middle of the spring. By means of its threaded rod, this safely pulls the spring together between the top plate on the car's body and a metal disc which is inserted at the bottom of the spring. Only then can the support be removed, and the lower wishbone withdrawn. If you do not have a lift and have to work with a jack, place a wooden block under the wishbone and slightly lower the jack. Once again, only remove the wooden block when the spring compressor is correctly positioned.

Conversely, when reassembling the parts, support the wishbone until the shock absorber has been refitted, and, in so doing, the spring travel has been restricted. As a matter of principle, some specialists use bushes made

On the Mercedes, an eccentric steering knuckle is joined to the wishbone at the top.

At the bottom, there is a sturdy threaded bolt.

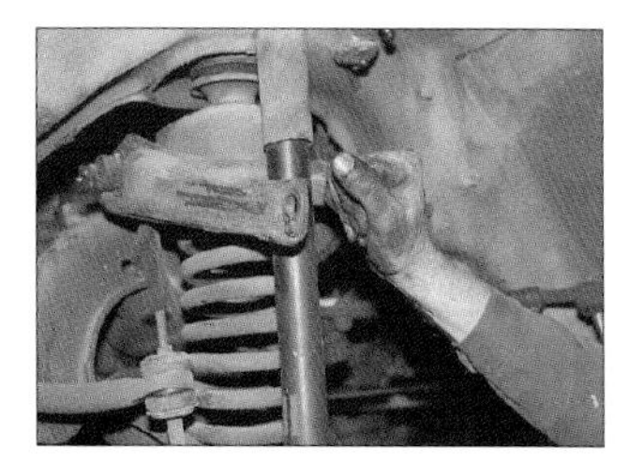
If things wobble here, it may only be that the fastening bolts have become loose.

The inner links on the 280 only needed some fresh grease.

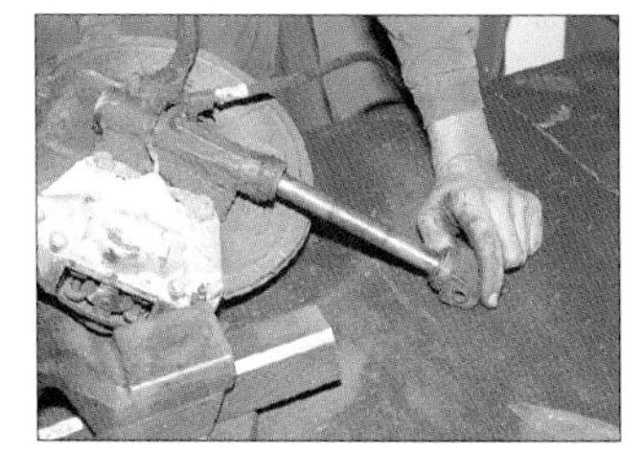
Here the kingpin is pulled upwards.

It is essential to note how the sealing rings have been fitted.

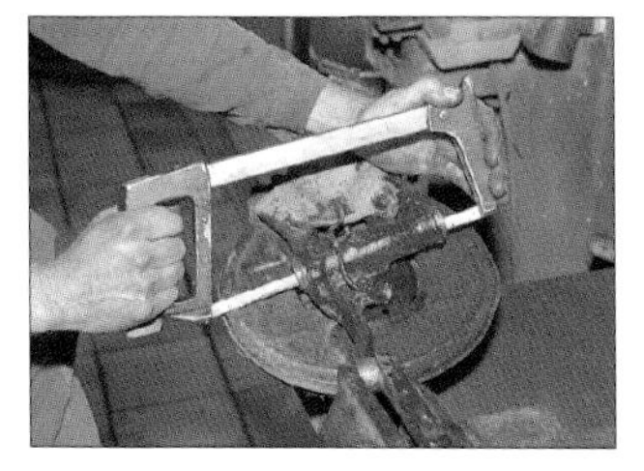
Cutting into them with a saw makes it easier to remove the old bushes.

A suitable hammer and punch will let you finish the job.

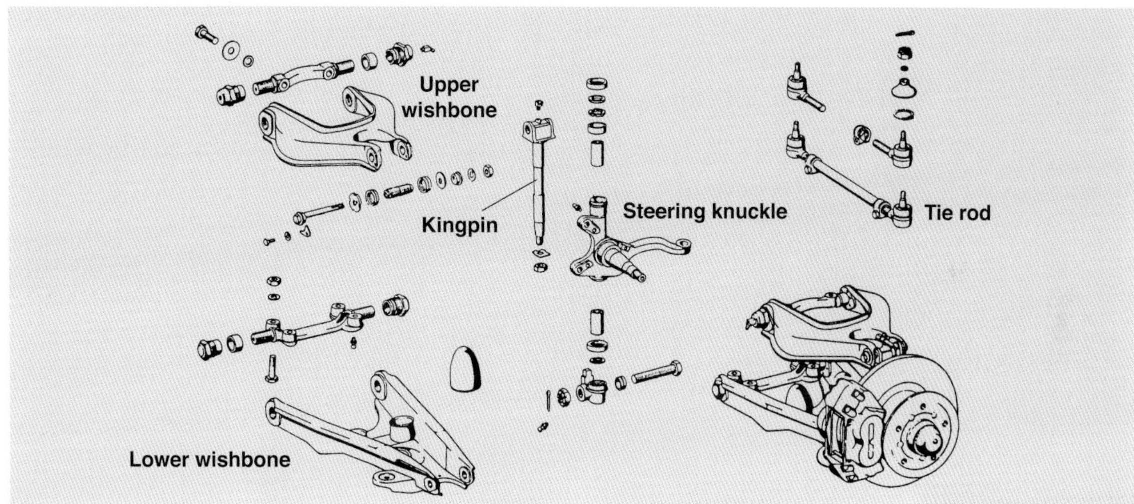

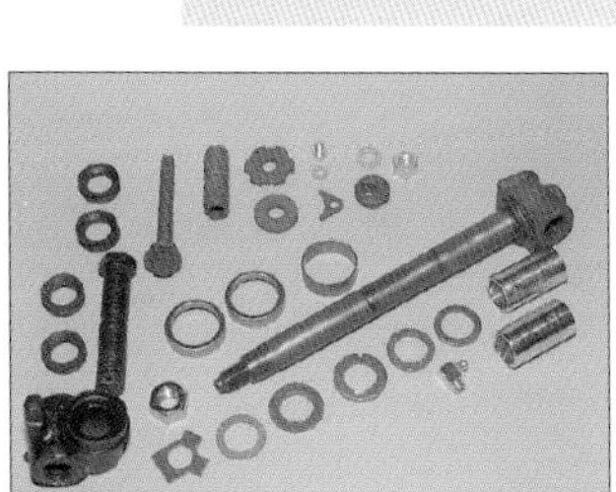
We're in luck – Mercedes offers a complete repair kit!

The new bushes can be pressed in using a vice.

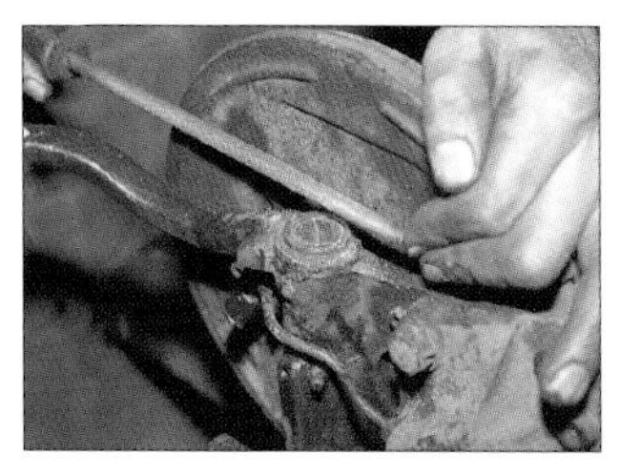
Two lubrication grooves must be cut with a file.

A reamer is used here to ensure the correct diameter.

When fitting the kingpin, pay particular attention to this groove.

There are three different spacers to adjust the amount of play.

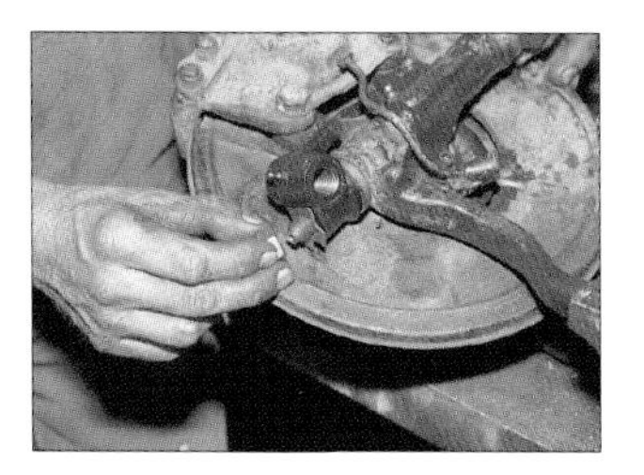
If the correct spacer is fitted, there should be about 1mm (0.04in) free play.

Faulty grease nipples can also be conveniently replaced.

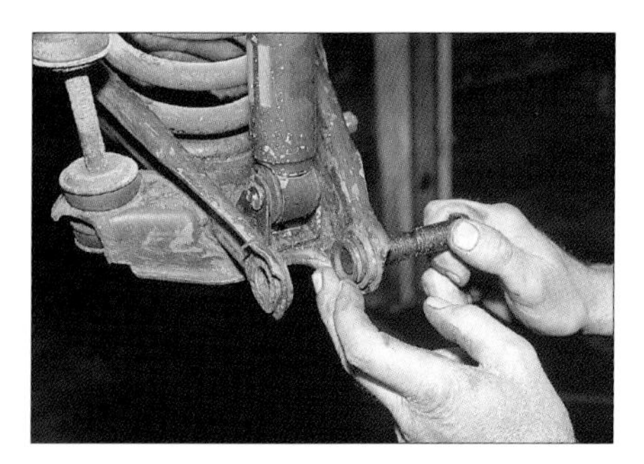

Just position the threaded bolt, then ...

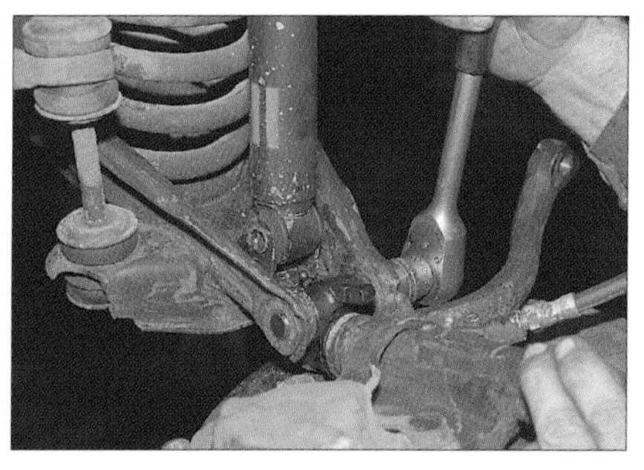

... insert both rubber rings together with the lower link and tighten everything up.

Next comes the eccentrically-mounted steering knuckle ...

... whose different components must later follow the correct order.

The camber angle can now be adjusted by means of the toothed wheel.

Tightening the nut at the bottom offsets the vertical play in the steering knuckle – and the job is complete!

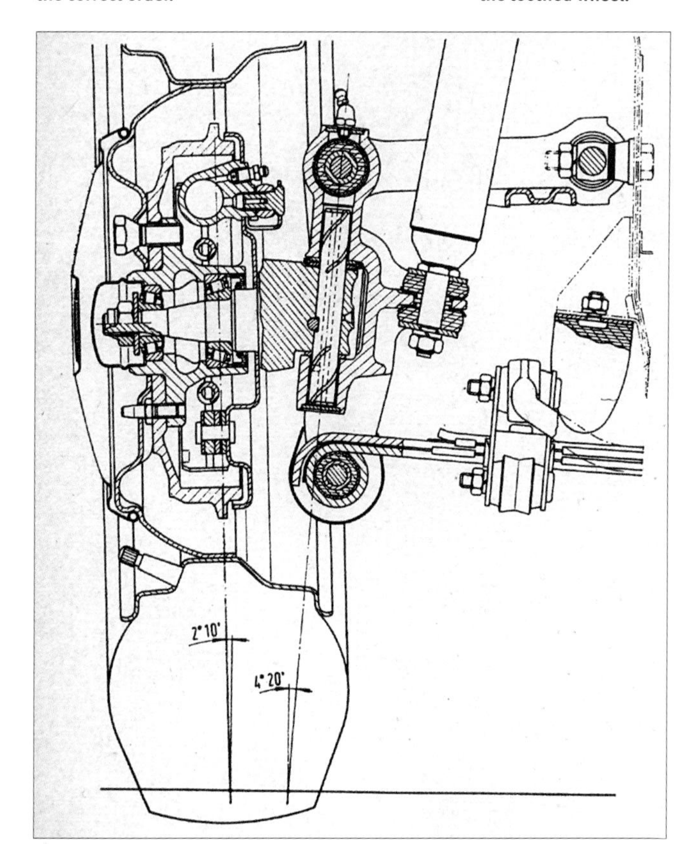

Another way of going about it: the kingpin on the Fiat 850 is a much more delicate affair.

from polyurethane – sold under the 'Superflex' brand name – when repairing axle assemblies. Experience has shown that this blue material is significantly longer-lasting than the traditional rubber components, and is very easy

Fiat also offers a complete repair kit for its 850. Interestingly, the lubrication grooves are not set into the bushes, but are milled directly into the kingpin itself ...

to press into place. Most experts use an hydraulic press to remove and refit the parts. Machined parts, mainly made from brass, are produced to fit, and serve as an adapter so that the punch does not press directly onto the new bush, with the risk of damaging it. Most bushes, however, can be removed and refitted using a vice with aluminium jaws, along with a suitable socket. The job on the TR6 was completed by fitting a new tie rod joint. Virtually the entire axle assembly was overhauled, leaving only the old anti-roll bar joint, as that proved to be in top condition. As some of the ball joints available today are of dubious quality, it's worth considering replacing the part with an original component in good condition. Check the rubber mounts of the anti-roll bar, as these can be changed at the same time.

The repair kit for the TR6 cost about •x450, to which should be added 10 hours' labour by the specialist. Doing the job yourself would save •x530-620 in labour charges, but you should reckon on it taking a day and a half if you are less experienced.

Turning now to the Mercedes, its front suspension also features a double wishbone on each side, but, in contrast to the TR6, it has a massive kingpin, which can be replaced when it wears out. It is typical of the suspension on the W108 that the inner links for the wishbones show minimal wear even after many years. In general, however, the kingpin, together with its bearing, and the outer wishbone, need to be replaced. This was exactly the case with our 280, which also exhibited one particular fault that is not unusual on this series of cars: the mount by which the upper wishbone is attached to the car's body had come unbolted. Our expert noted: "We come up against this very regularly. Even though there are sturdy reinforcement plates, the bolts work loose and cause a rattling sound from the front suspension, which is often confused with much more serious damage." With a bit of luck, the overhaul of the front axle may come down to tightening a few bolts. With the W108 as well, the lower wishbone had to be supported before the steering knuckle was removed, as the shock absorber also restricts the rebound travel. For other makes of car, a quick glance at the repair manual can be a godsend and prevent a spring suddenly flying past your ears! As the shock absorbers and their mounts had been replaced on the 280 not long before, we dispensed with their removal. It's much easier to remove the bushes for the kingpin if you first cut into them slightly lengthwise (don't cut them all the way through!). After cleaning the assembly, the new bushes could be pressed home. As not everyone has a press, we have shown this stage using a vice. Under no circumstances should the bushes be pressed in at an angle.

The bushes were adjusted to fit the kingpin, using an adjustable reamer with tapered guides on both sides, so that it moved freely and with no perceptible play. If you are heavy-handed using the reamer, it's easy to go too far and have to replace the bushes once again.

The lower joint has a conical seat on the kingpin. First, it is attached with one of three differently sized spacers with about 1mm (0.04in) free play. Only when the steering knuckle has been completely refitted to both wishbones is the vertical play adjusted to zero by tightening the nut. If it pivots easily with no vertical play, everything is all right. If some play remains, the spacers must be exchanged for the next larger size. The thinner spacers come into play when the kingpin can only be turned with difficulty. The pin which connects the upper wishbone to the kingpin operates eccentrically in relation to the camber adjustment.

The repair kit shown costs about •x710 from Mercedes. If you have all the work done by a specialist, you should allow for a day in the workshop. Doing it yourself can save over •x620.

Axle assemblies with MacPherson struts are completely different in their construction from the Triumph and Mercedes items. This design originated with the Ford engineer Earle S MacPherson in the late 1940s. He combined the shock absorber, spring and wheel mount in a single pivoting unit. At the top, the strut is

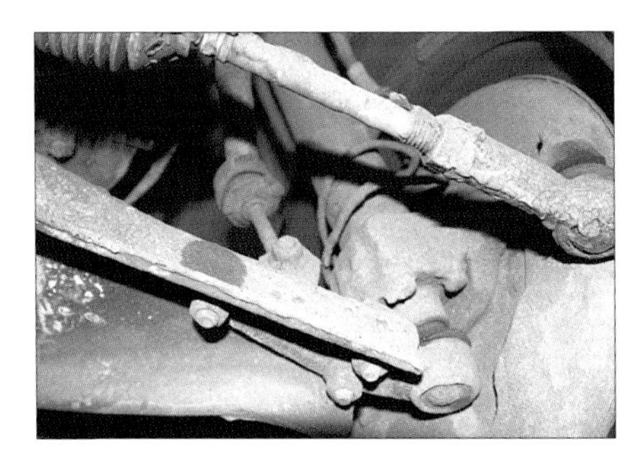

Rattling away: the supporting links for the MacPherson strut on this Volvo 245 are broken.

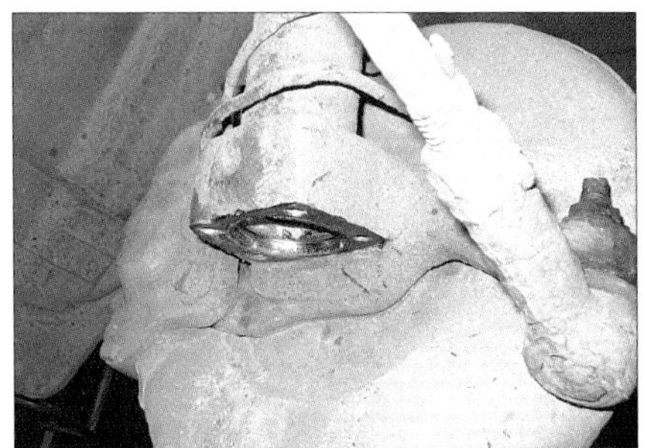

No need for a press: the mounting for the link is bolted in place, and the strut also stays put.

New bolts are thoughtfully included with the replacement supporting links.

An easy job: anyone can tighten a bolt. Unfortunately, the new bolts are a bit too long.

The original parts have not been cleaned up and the link can be bolted in exactly the right position.

An essential tool: never take suspension struts apart without using a spring compressor, as they can be very dangerous otherwise.

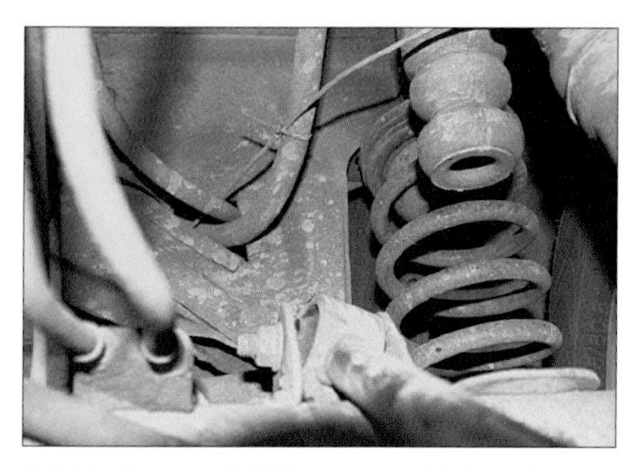

Quickly diagnosed: if the torque rod is bent in this way, the damage is obvious.

Surprisingly, although the part was almost new, it was nonetheless broken. Clearly, it had been incorrectly fitted.

Strength required: although the bush was quite new, it was firmly rusted in place.

At an angle: using a vice, the new part could not be accurately pressed in.

On the hydraulic press, however, there was no problem.

The job was harder than expected, but successful in the end: the new bush was fitted.

Proper control restored: the Volvo's suspension could now do its job again without any untoward noises.

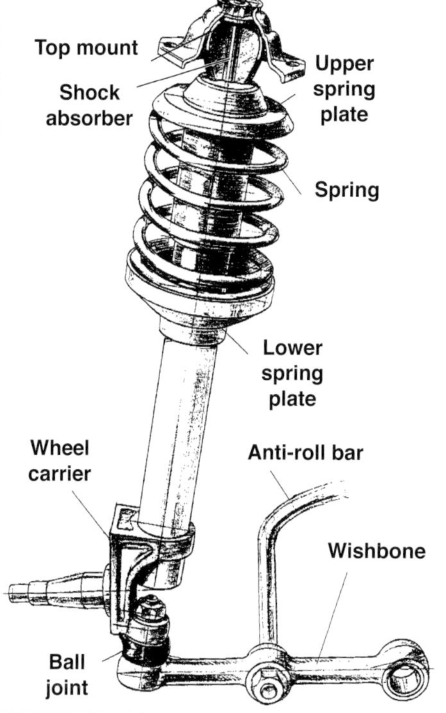

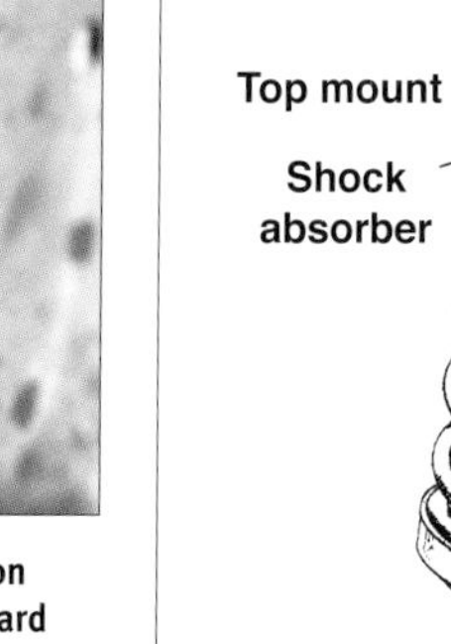

bolted into a supporting mount inside the wheelarch, while, at the bottom, it's connected via a ball joint to a wishbone.

The first European cars were fitted with this design of strut in the 1960s. Damage typically occurs to the top mount and the bottom ball joint, and sometimes to the wishbone links as well. Changing the shock absorber cartridge is a laborious undertaking, and virtually impossible without a good spring compressor. The Volvo 245, whose basic design dates back to 1965, has MacPherson struts exactly like these. On our model, the lower supporting links were in need of replacement. The top mounts and shock absorber assemblies were in good condition, sparing us an unpleasant job. On late-model 245s, there's no need to strip down the strut when changing the links, it's enough to loosen the shock absorber assembly. The links on the Volvo are bolted to the wishbone, so that it's only necessary to take off the anti-roll bar and the wishbone. Where the link is attached to the strut, however, is a critical point: an adapter plate is used to fix the bolt. This is attached to the strut housing via four M8 bolts in an area where it is exposed to road spray. If the bolts break off, they will need to be drilled out; fortunately, ours came undone without difficulty. The new supporting link came with all the necessary screws. Note that the specialist chose not to clean the wishbone, as the mark left in the dirt made it easier to see the position of the old link. If you bolt on the new part in exactly the same position, the front suspension will be set up more or less correctly. If a precise optical alignment check cannot be carried out on the spot, at least it will be safe to drive the car to a wheel alignment specialist.

The Volvo had its worst problems at the rear. The entire 200 series has a very sturdy live axle, which is bolted to the body by way of the trailing arms and shock absorbers. To reduce roll under different load conditions, it also has two torque rods and a Panhard rod which acts as a transverse control arm. If the bushes here are faulty,

The rear suspension as installed.
1 Mounting bracket
2 Torque rod
3 Mounting bracket
4 Rubber bump-stop
5 Rear spring
6 Mounting bracket
7 Anti-roll bar
8 Rear crossmember
9 Upper shock absorber mount
10 Disc
11 Rubber bump-stop
12 Mounting bracket
13 Lower spring mounting bolt
14 Disc
15 Suspension arm
16 Shock absorber
17 Lower shock absorber mounting
18 Front mounting for torque rod
19 Suspension arm bushes

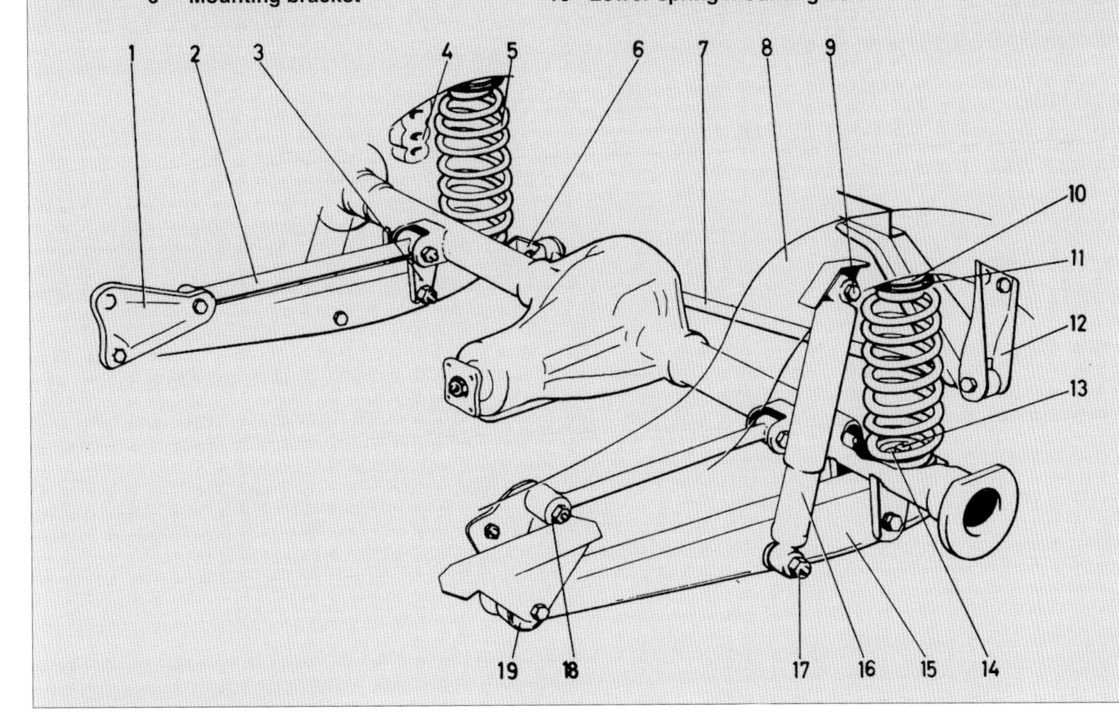

Supporting links which have been press-fitted cannot be dismantled using a vice.

Only an hydraulic press will do the job ...

... together with suitable adapters. The old link on the Beetle was quickly removed.

Rust and dirt should be cleaned off the strut eye.

The link has grooves which provide protection against twisting.

Before it is pressed in, the collar is protected with a sleeve, to which ...

... further slow pressure is applied. The link ...

... now gradually slides into the strut eye, under a pressure of 9.5 tons.

The new part has been fitted and is correctly positioned.

Not ideal: the new link has a flatter end, which will collect dirt.

Wear the right protective headgear: if something buckles up on the press, it can be dangerous.

An effective solution: on the VW Microbus and Beetle, the camber is adjusted by means of eccentric bushes ...

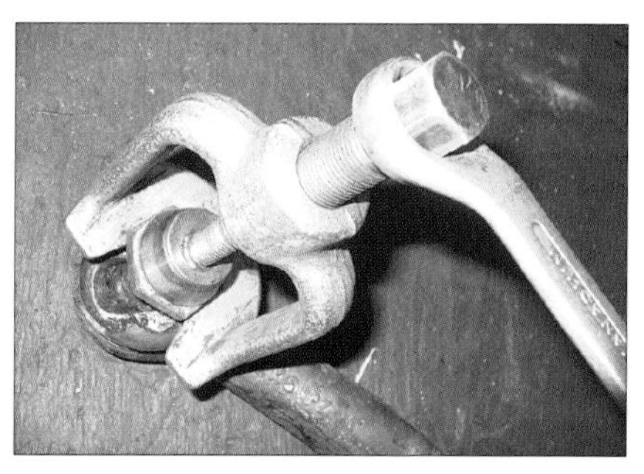

... which always require a suitable tool to remove them.

Cotter pins are hardly ever fitted; nowadays, nuts are self-locking.

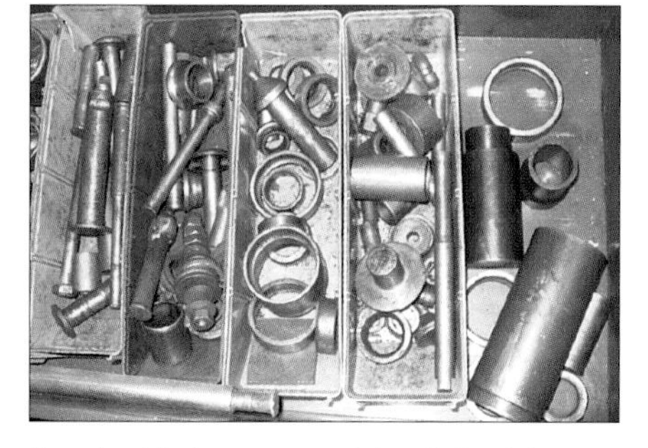

A mechanic's dream: a complete set of metal sleeves to use on the press.

The Volvo 245 once again: here the inner sleeve has been completely torn out of the rubber.

Sure to make a racket: these rubber supports for the anti-roll bar mounting are hopelessly cracked.

Driven to the test centre and immediately failed: this knuckle ...

... was on the point of falling out. The poor-quality rubber gaiter had already crumbled away.

Another example of poor-quality parts: these silent blocks lasted barely three years.

the car can develop a dangerous tendency to steer of its own accord during the weight transfer which occurs when leaving a corner (this applies, by the way, to all designs of rear suspension).

As there were knocking sounds coming from the Volvo's suspension, but it drove okay, the upper torque rods were the likely culprits, and would be fitted with new silent blocks.

When the parts were removed, a classic mistake in assembly became apparent: as a rule, parts which are bolted to the chassis using bushes should be tightened when the springs are compressed; that is, when the car is standing on the ground. Plainly, this had been overlooked in this case: the arms on the Volvo were practically new, but were already completely worn out! If, for convenience sake, you tighten the bolts with the car on the lift when the wheels are not taking any of the weight, you'll end up fitting the inner bushing in the wrong position. Later, when the springs are compressed, the bushing will be twisted against the rubber, and will come away in next to no time. Which was exactly what had happened here. The fitter suggested that, "With old cars like this 245, it is often the apprentices in garages who work on them, nobody really wants to work on old cars like these any more. The work isn't checked afterwards, and so mistakes fitting the parts occur. The quality of the bushes wasn't an issue, as they were original Volvo parts."

Curiously enough, it also proved difficult to press out the broken bushes on the torque rods, although they had only been in use on the car for two years, and showed hardly any signs of corrosion. The force required was too great for the vice, so we switched over to the hydraulic press. It took a pressure of no less than eight tons to force out the old bushes. Even the new bushes required almost as much pressure to fit them. After the torque rods had been refitted, the knocking sounds disappeared. The complete job cost about ●x180 for the replacement parts and ●x360 in labour. It's possible to buy the torque rods as a complete kit, so that a home mechanic could do the job on the Volvo, even without access to an hydraulic press.

Now for the Beetle. The VW Beetle's front suspension

design was already obsolete by the 1970s, but its supporting links were still up-to-date, and remain so today. Press-fitting these is scarcely any different from any other car.

There are few options open to the amateur mechanic: press-fitting new links is not a job that can be done using a vice. You can, though, remove the old parts: it's possible to remove the upper sealing cap and drive out the bolt complete with the bearing shells. Next, the link housing must be cut using a metal saw until the link loses its tension and can be removed through the eye in the strut. The whole job is just as tedious as it sounds, and fitting the new links has to be done with a press. Removing the old components also takes a matter of minutes using a press. Before this, a puller must be used to take apart the eccentric bush for the camber adjustment for the upper strut. This will be reused. When pressing out the link, the strut eye must be backed up with a washer, and, on the reverse side of the link, a suitable tubular sleeve is pushed through. The new link has a fixed mounting position, which is marked with two notches. Once any rust has been cleaned off the strut eye, the link is positioned and pressed in using an appropriate sleeve. Note that full-face protection should be worn whilst doing this. If a metal component breaks or slips away under a pressure of several tons, it can be lethal.

Finally, some general advice. First, any bolts which originally came with reinforcement plates or cotter pins should always be re-attached in the same way, and, in principle, the plates and pins should be replaced. This also applies to self-locking nuts.

Second, pay attention to the thread pitch: most car parts use a fine pitch. Bolts in linkages where rubber bushes are fitted with inner and outer metal sleeves should only be tightened after the springs have been compressed and released several times. This also applies to cars where the workshop manual does not specify this: this will prevent the rubber twisting against the sleeve.

Third, any parts which are completely intact should be reused. The quality of tie rods, supporting links and steering knuckle repair kits from many manufacturers has, unfortunately, declined. Whenever possible, an original component from old stock should be your first choice for repairs.

Finally, after overhauling an axle assembly, the camber, tracking and castor should be checked using suitable test equipment. Tyre suppliers frequently offer inexpensive alignment checks. In any event, take the repair manual with you, as firms which use computerised alignment equipment no longer have the setting data for older cars. For some cars, the basic settings in the repair manual are good enough, and no further adjustment is needed; but don't rely on this! Drive very carefully on your way to this final check, as dramatic changes to the suspension geometry may have taken place during the repair work. You will only experience the new feel of the car after the camber, tracking and castor are all correct once more.

Pitstop

Cause for celebration: the materials for the home-made polyurethane bushes cost just ●x1.35.

For many popular classics, complete sets of polyurethane suspension bushes have been offered for many years. The new polyurethane parts should be more durable than the original components, and are sometimes even available in different degrees of hardness. But what can you do if plastic bushes aren't available for your classic? Two-component polyurethane means you can mould new suspension bushes yourself at home.

Bushes such as these usually have a steel inner and outer sleeves. Only in exceptional cases will you find tubing in metalwork shops with the exact diameter and thickness that you require; these sleeves will therefore need to be produced on a lathe. Even if your bush does not have an outer sleeve, you'll need to make one to serve as a mould (and then treat it with silicon spray as a separating agent).

Two-component polyurethane is mixed in a ratio of one part polyurethane to 0.35 parts hardener, then slowly and evenly stirred for at least four minutes (to prevent air pockets); it can then be used after 20 minutes. After the mould has been cast, the material takes four hours to set, and you should wait another eight days before fitting the part.

810g (29oz) of polyurethane from Weico costs about ●x45. To produce a simple bush for a control arm requires about 15-20g (0.6-0.7oz), so making a bush such as this will cost you just ●x1.35, plus any costs you may incur for the inner and outer sleeves. The maximum possible thickness is 75mm (3in).

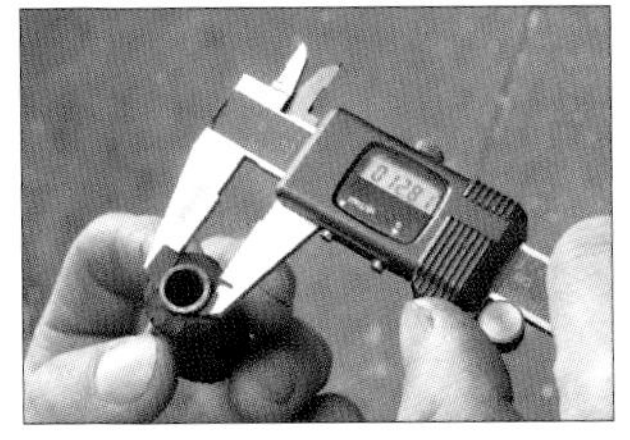

Measuring up: first, the inner and outer sleeves are made, based on the old part.

The newly produced sleeves are fixed onto a magnetic plate and will serve as moulds.

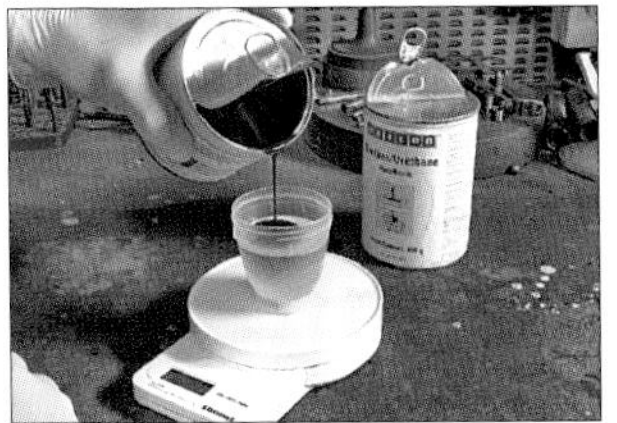

The polyurethane and hardener are mixed in a ratio of one part polyurethane to 0.35 parts hardener and thoroughly stirred.

The material can be used after about 20 minutes. At 20°C (68°F), it takes around four hours to dry.

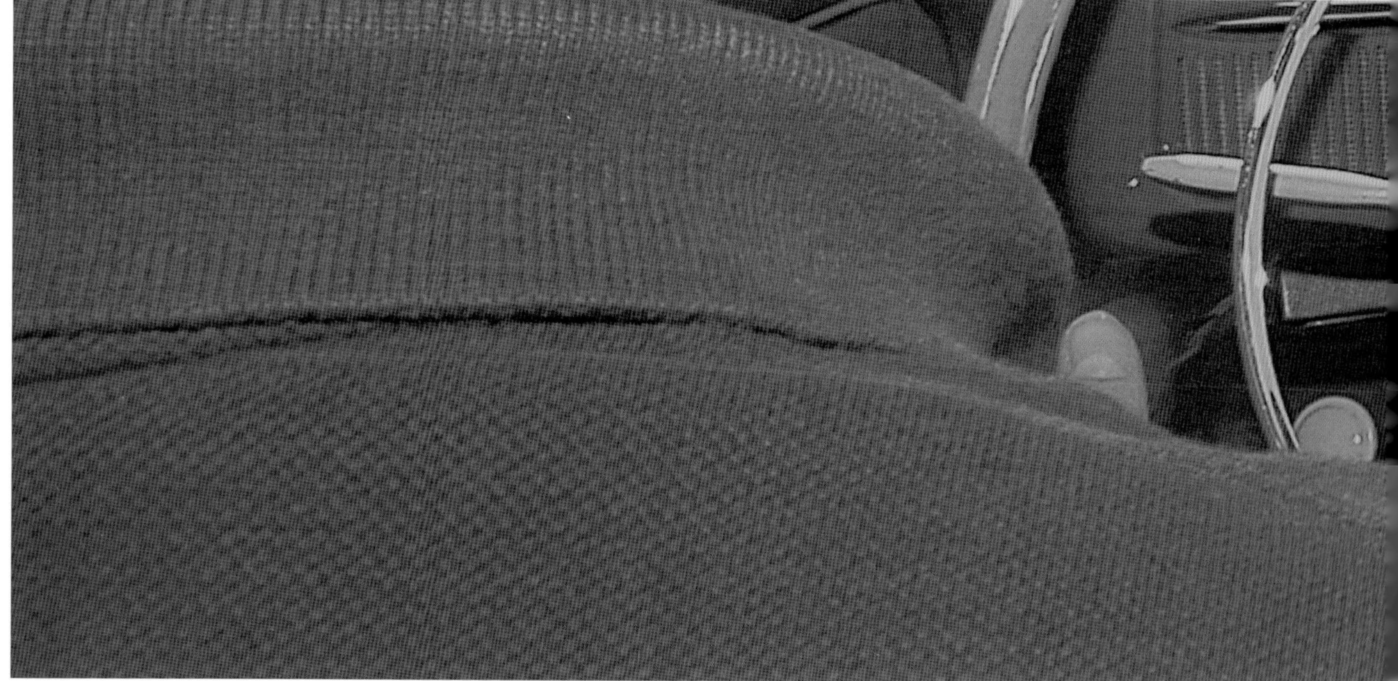

Steering

Different types of steering

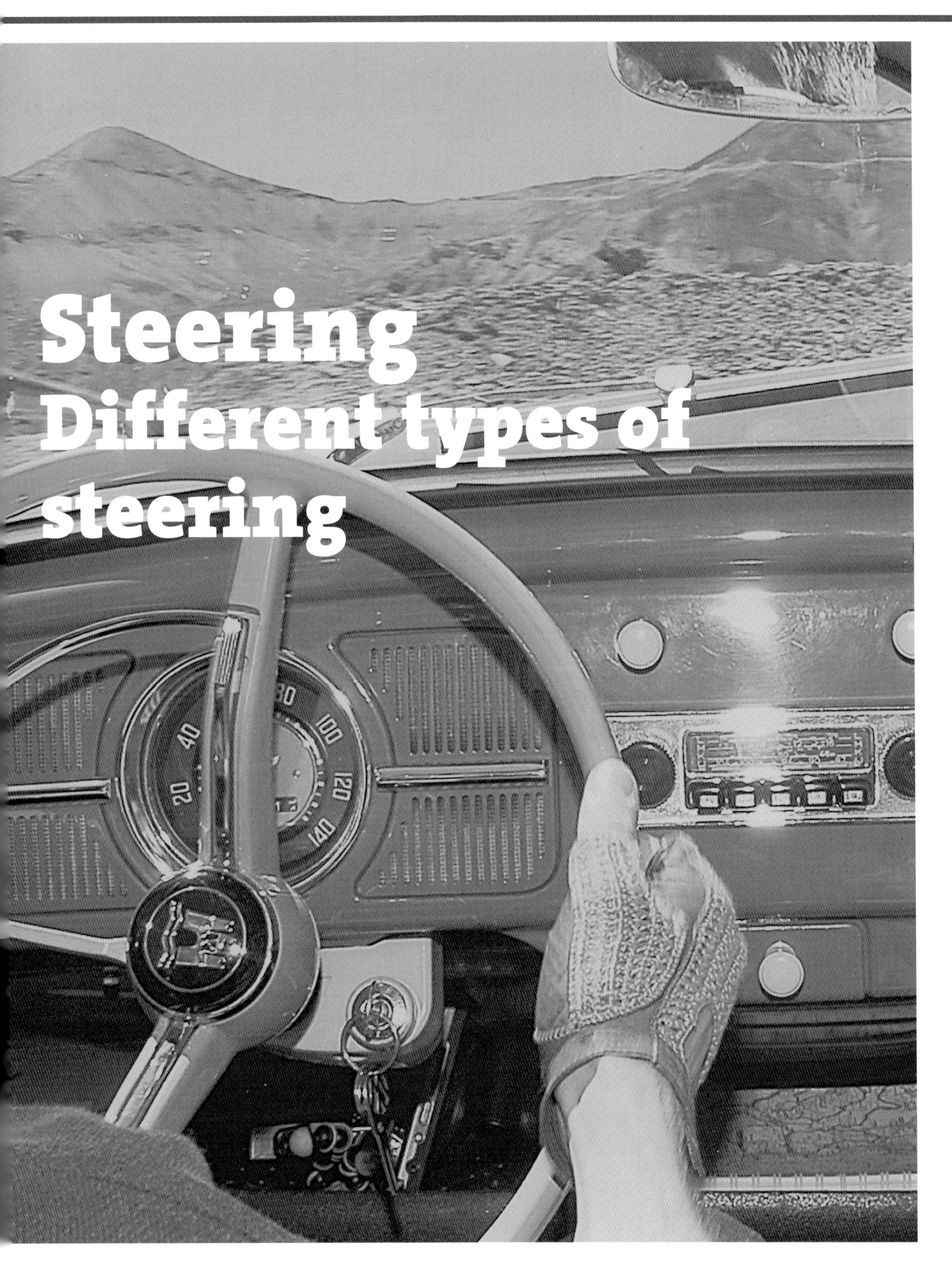

Hardly any system connects the driver so directly to his or her car than its steering. A variety of mechanical solutions exist to make the car change direction as the driver wishes.

Nearly all common types of steering follow the so-called 'Ackermann' principle, invented by the carriage builder from Munich Georg Lankensperger in the early 19th century. The principle's distinguishing feature is that the axle itself does not turn (unlike the fifth-wheel steering used on articulated lorries); rather, the wheels run on so-called 'steering knuckles,' which turn around a pivot; the kingpin. Different designs of steering gear then do the job of converting the rotary motion of the steering wheel into a swivelling movement of the steering knuckles, with the help of a system of steering arms and tie rods.

The most important designs of steering include worm and nut and recirculating ball steering. The steering column, which turns directly with the steering wheel, may also be referred to as a spindle, as there is a spindle-shaped thread at its end. On this thread is a nut, which itself cannot turn and therefore moves up or down the spindle as the spindle rotates, depending on the direction in which it turns, as is quite normal for any nut and bolt. The steering arm is connected to this nut by means of a special bearing. Through the movement of the steering shaft, which is in turn a result of the up-and-down movement of the nut, the steering arm swivels; this movement is then transferred to the wheels via the steering linkages. The whole system is very reliable and long-lasting in operation, while the relatively modest load on the thread ensures that the nut moves quite easily. But the mechanism is subject to friction and wear, so it is important to keep it well lubricated.

Inside a worm and sector steering system there is a steering worm, which turns in response to the steering wheel. A worm gear mounted at right angles to the steering worm engages with it (a cog segment is also sufficient). The rotary motion causes the worm gear to move to the left or right. The steering arm, which is solidly connected to the worm shaft, converts this rotary motion into a swivelling motion, and transfers it to the steering linkage and from there to the wheels.

Worm and roller steering systems are more recent. In this case, the so-called 'globoid' worm tapers towards the middle and causes a steering shaft to move, which acts

With worm and sector steering (above) and worm and roller steering (right) ...

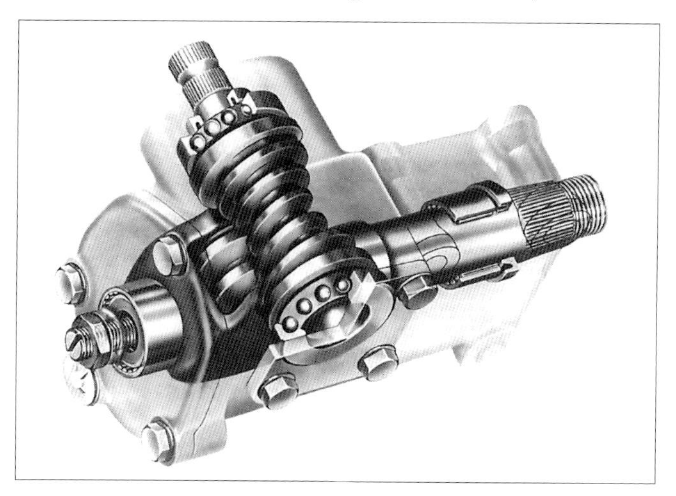

... a worm thread, which turns with the steering wheel, acts on a worm gear or roller.

The Ross-type steering is also a form of worm and sector steering ...

... but here the worm thread at the end of the steering shaft moves a lever.

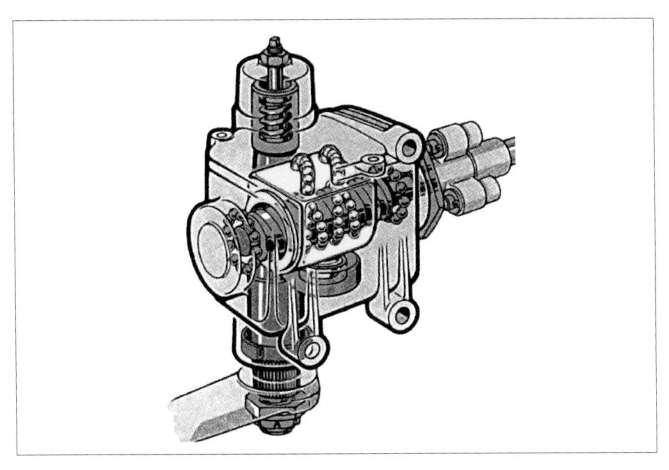

Recirculating ball systems work on the principle of screw and nut steering. The ball bearings ...

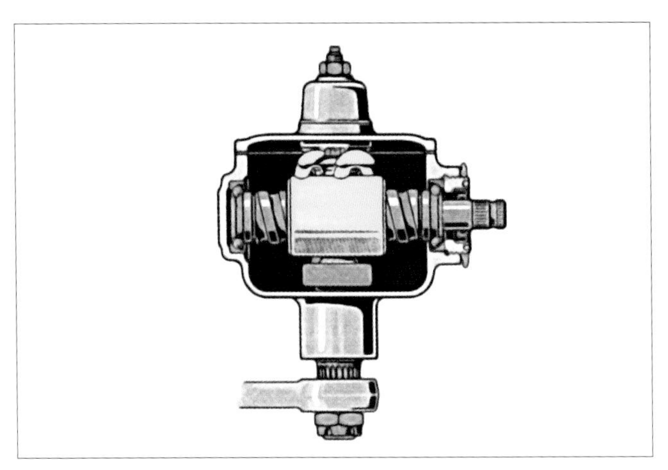

... ensure that there is less friction between the steering shaft and nut.

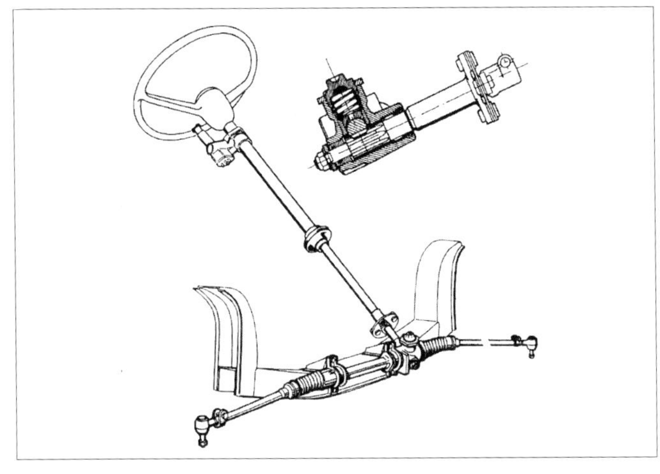

The rack and pinion steering mechanism is mounted at right angles to the direction of travel between two tie rods.

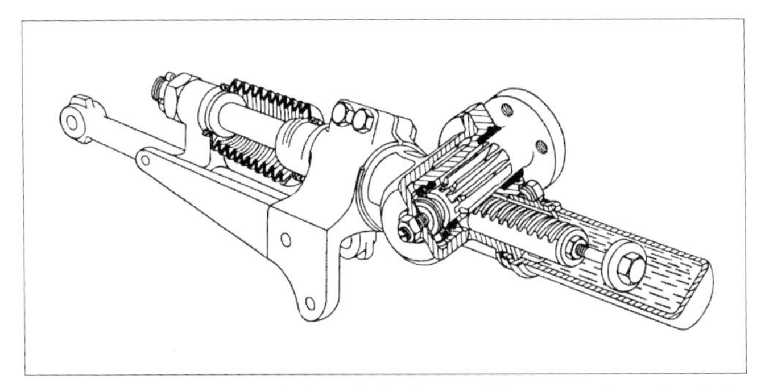

Rack and pinion steering: simple and effective. A pinion moves the rack, which then directly moves the tie rods. Depending on the setup, the rack and pinion may have straight-cut or helical teeth.

as a toothed roller with two or three gear teeth. This runs in roller bearings carried in a fork, which is located on the steering shaft. As the steering worm turns, the roller swivels and so causes the steering shaft to turn, and with it the steering arm. ZF's Gemmer steering is very similar to a worm and roller system. Cam and peg (or Ross-type) steering also works on the same principles, but the worm shaft is replaced by a single (or sometimes two) cam lever(s) which engage with the worm drive. This cam lever is mounted on a cross-shaft lever, which in turn directly actions the steering shaft. The advantage of this design is that the cam is mounted on ball bearings, which reduce friction to a minimum. In the event of wear, the cam lever can be made to engage better by simply undoing a locknut and turning an adjustment screw.

Rack and pinion steering functions in a simple but precise manner: a gearwheel, a rack and a single housing for both parts, and that is all for the most important components. The gearwheel is turned by the steering wheel and engages directly with the rack, making it move to the left or right. The rack is often pressed against the pinion with a spring-loaded thrust piece, which absorbs any play in the mechanism – a form of self-adjustment that is as simple as it is effective. In addition, the rack is frequently slightly curved, so that, even when heavy wear occurs in its middle section (which is usually most stressed), it can be adjusted to turn without sticking. Often, the rack is positioned in the centre of the separate tie rods, so that it acts almost directly via the steering arm on the steering knuckles – it could scarcely be easier! Since this design is not only effective, but compact and inexpensive to produce, most modern cars now use rack and pinion steering.

Rack and pinion steering failed to catch on earlier because of limitations in the steering ratios that were possible. Originally, the direct nature of this type of steering created the problem that uneven road surfaces could be felt through the steering wheel as unpleasant jolts. Rubber bushes and flexible joints in the steering column have remedied this weakness, albeit with a slight loss in steering precision.

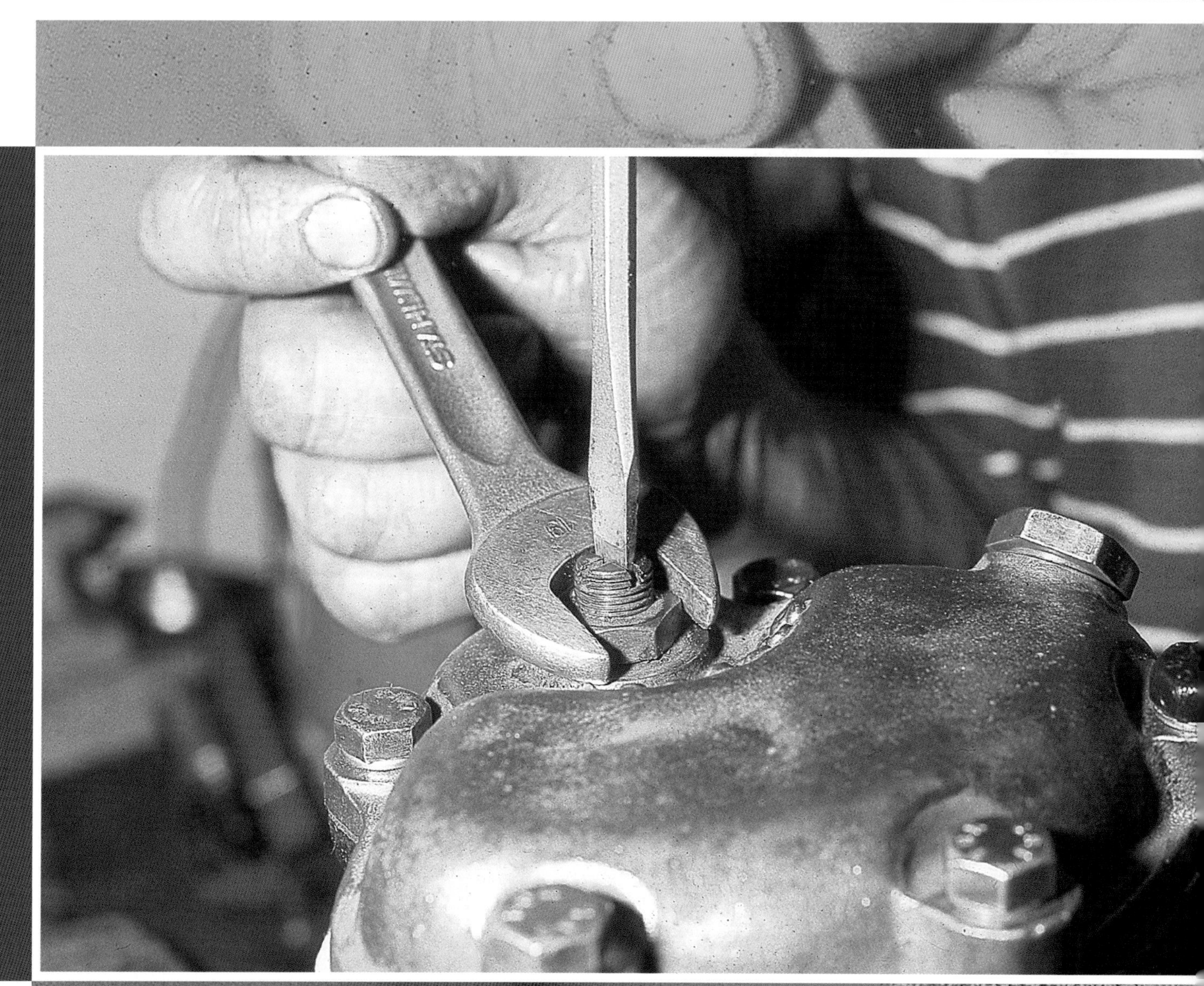

Steering
Overhauling steering systems

Steering systems, for the most part, go about their work without complaint. They usually operate so unobtrusively that they are often neglected. What should you do if they play up?

There can be many causes leading to their demise. And it's always then, when it's already too late, that well-meant advice will rain down from every side. That's when people will talk about adjusting the play in the mechanism, about lubricating it regularly, and about all the mistakes you can make during maintenance work.

If the steering is no longer working perfectly, the first thing is to pinpoint the cause of the symptoms. That is most easily done with the car on a lift, so that all the steering components can be moved freely. Have someone turn the steering wheel while you get underneath the car. An increase in free play does not usually arise in the steering gear itself, but in the connecting elements, such as the trackrod ends, idler arm or steering knuckles. That's why these components are checked during every safety inspection.

Cutaway model of a VW spindle-type steering mechanism as installed on the front axle.

Steering forces are usually transmitted via worm and sector, recirculating ball, or rack and pinion set-ups. In all three main types, a thread is located at the end of the steering column, which is connected to the upper end of the steering shaft or drop arm. It works like a screw and nut: if you turn the screw to the left, the nut moves upwards, if you turn it to the right, the nut moves down. In order to keep the friction between the 'screw' and 'nut' to a minimum, in the case of recirculating ball steering, ball bearings are placed between them, with a simple metal channel to connect the lower and upper ends of the thread and stop the ball bearings simply rolling out. Thanks to the low degree of friction, the system works very smoothly, with hardly any wear, and requiring virtually no maintenance. Only the level of the oil in the steering box needs to be checked regularly: SAE 90 or SAE 90 EP gearbox oil should always be used. The disadvantage of this system is that if just one of the ball bearings is damaged because of a sudden heavy loading (such as can occur when a front wheel drops into a pothole) or insufficient lubrication, in next to no time the entire mechanism will be ready for the scrap heap. The remnants of a ball bearing which has disintegrated will be distributed throughout the mechanism, damaging the track which the ball bearings follow. A clattering noise in the steering is a sure indication of this. At this initial stage, try to replace the old ball bearings, assuming the track itself has not been significantly damaged.

A worm and sector steering system is much easier to envisage. As its name implies, at the end of the drop arm there is a worm gear, which engages a sector shaft which has teeth cut around its surface and is tapered in the middle. There is scarcely any friction between the intermeshing gears with this type of mechanism, as the worm gear turns with each rotation of the shaft. The free play between the worm gear and the shaft can be adjusted by means of a screw in the steering box. This screw presses down on the drop arm from above, forcing the worm gear deeper into the toothed mechanism of the shaft.

Time and again, the same mistake is made when adjusting worm and sector steering. Since a car is most often driven in a straight line, most wear occurs in the

VW produced this unusual spindle-type steering mechanism, in which the spindle causes a spherical cup to move: this rests inside a half-shell in the steering shaft and so turns the drop arm.

Watch out for small parts when opening up the steering box.

On the cover: the adjustment screw and the opening to check the oil level.

With the cover removed, the workings can be seen.

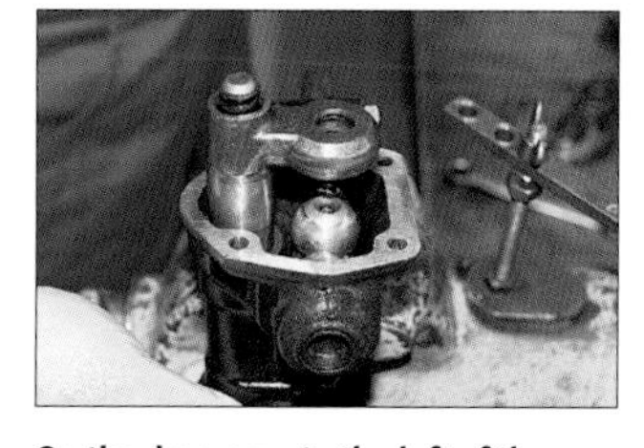
On the drop arm, to the left of the picture, the thrust piece and spring are ready to be refitted.

You can remove the steering shaft, leaving the spherical cup in place.

This old steering mechanism is filled with gearbox oil (in this case SAE 90).

Traces of wear can be seen on the spherical cup, which may cause the mechanism to stiffen.

That's all there is to it: this VW steering gear was manufactured before the war.

One advantage regarding the supply of spare parts for this steering mechanism is that ...

... you can make paper gaskets yourself. Tailor-made plastic seals were only introduced later.

central position of the steering. If you adjust the steering here such that all the free play is taken up, it will be hard to turn when you apply more lock. Adjusting old worm and sector steering mechanisms is often a compromise between an acceptable degree of play in the central position and smooth operation when cornering.

Volkswagen produced an unusual spindle-based steering set-up until 1961. In this, the steering spindle causes a spherical cup, which rests inside a half-shell in the steering shaft, to move. Volkswagen built the steering mechanism, but used steering boxes supplied by the Georg Fischer foundry (hence the abbreviation 'GF' stamped on the castings). Before the assembly shown here was removed, the following symptoms were observed: the Beetle's steering had been abnormally heavy for some time in normal driving; substantial effort was required to return the wheel to the central position; even the attempt to find a compromise between acceptable play and smooth operation was to no avail (when the steering was set up to avoid becoming heavy, there was a full quarter turn of free play). When the steering was stripped down, the spherical cup displayed ridges caused by wear. Since no replacements were available, a new spherical cup had to be made.

It's sometimes possible that a temporary remedy can be effected by thoroughly cleaning and deburring the spherical cup, and applying modern grease instead of the gearbox oil. This is not a good long term solution, though.

Another option sometimes suggested is to replace worn parts using steering assemblies from cars which have been scrapped. These are often just as badly worn, however, so this is not recommended. A more reliable solution would be to replace the worn set-up with modern components, such as those fitted in Mexico until 2003. These are available quite cheaply, although a new drop arm and new trackrods may also be needed, as parts for Beetles from different model years are not always interchangeable.

The finger-type steering gear is not dissimilar to the spindle type. In this instance, a conical steering 'finger' engages with the thread of the steering shaft. The finger is located in tapered rollers (so it can be adjusted in relation to the drop arm), but the degree of friction with the shaft is relatively high. Finger-type steering gear can be adjusted via a screw on the outside, which presses down on the drop arm. All the pressure must be taken up by the two tapered roller bearings of the steering finger, which, understandably, makes these parts the most susceptible to wear. A sudden increase in loading, as when driving over a kerb, can damage the taper of the steering finger, with the result that it no longer turns as it should. The spot which has been levelled off then engages more and more frequently with the thread of the steering shaft, and jolts over its upper surface rather than rolling smoothly along it. The rate of wear increases enormously, increasingly affecting the shaft as well. It's also not uncommon for the taper to be flattened as a result of amateur attempts to make adjustments, particularly when the setting screw is tightened with too much force in the central position. That can be enough to flatten the taper on two sides; the reduced play when cornering does the rest.

The pictures which follow show how a finger-type steering mechanism is overhauled, taking a Mercedes 170 V as an example.

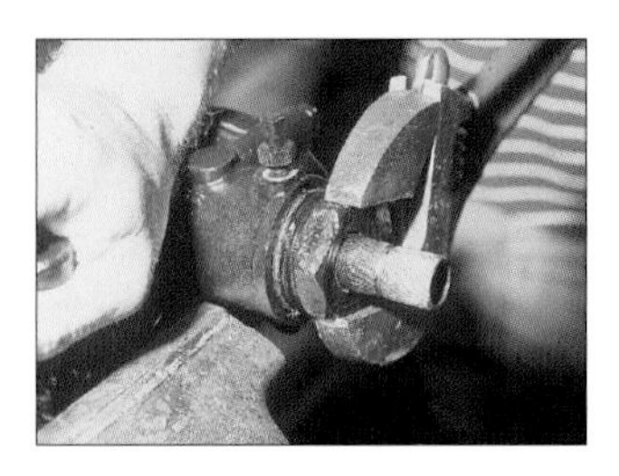

ZF spindle-type steering: release the safety catch (top) and retaining nut.

The drop arm can be removed from under the cover.

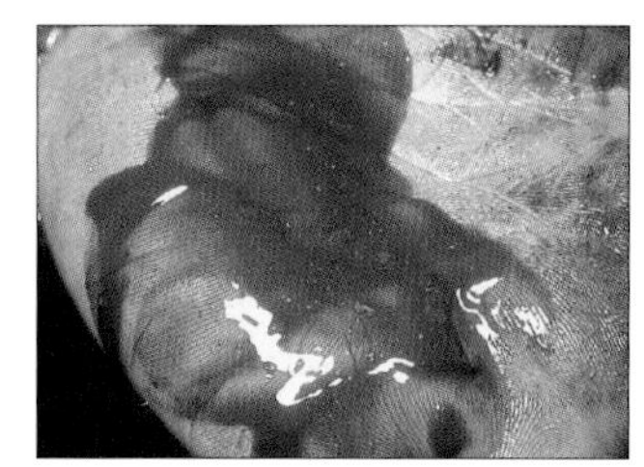

Traces of wear and swarf can be found in the oil.

Then pull out the spindle and inspect the casing for cracks.

After cold-cleaning the spindle ask yourself: can it be reused?

Next lever off the ring which retains the spindle bearings.

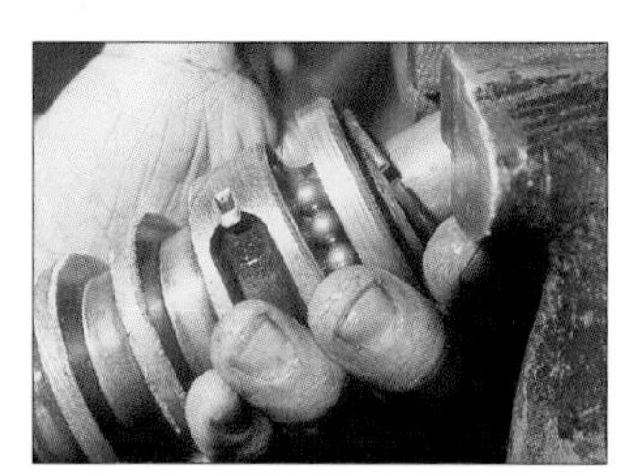

The upper set of bearings is opened up in the same way.

Scratched bearing surfaces make the spindle unusable.

In this case, the ball races were broken.

Remove the nut, and the steering finger, on ball bearings, is exposed.

A defective roller has caused the arm to reach its wear limit.

Replacements for faulty rollers like these are generally available.

Patience and plenty of grease are needed when inserting new rollers ...

... which can then be placed in the arm, which has been reground.

After this, the steering 'finger' can be carefully pressed home.

The ball bearings and ball race are also replaced.

The lightly greased drop arm is put back in place.

Modern, pressure-resistant bearing grease, replaces gearbox oil.

The copper seal rings should not be forgotten, or grease will seep out.

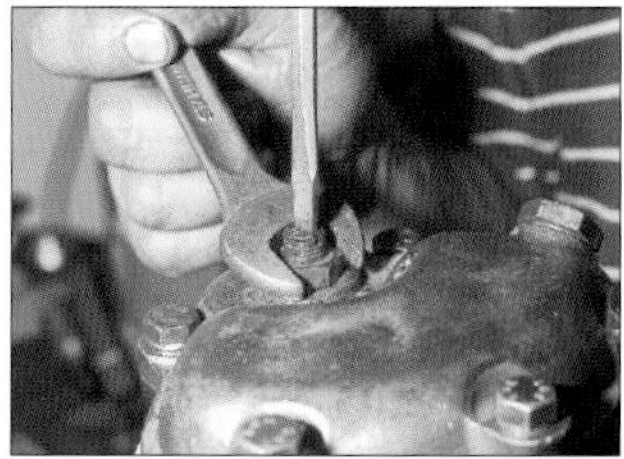

Adjustment is a compromise between smooth operation and free play.

Daimler-Benz obtained these steering mechanisms from ZF (Zahnradfabrik) in Friedrichshafen. The company has been supplying steering gear since 1932, and can still provide service backup and, if necessary, replacement steering parts for numerous cars which have since become classics.

The situation is considerably more difficult with products from other manufacturers, which are generally only available through their dealer network. When they no longer have any stock, and good alternatives do not exist, specialists with a lathe and a good feel for the job are much in demand. Rack and pinion systems occupy a special place among steering mechanisms. We looked over the shoulder of the technicians at ZF's branch in Holzwickede

Fit for scrap: an accident or driving over a kerb have caused chips in the edge of the teeth.

Fit for scrap: as a result of faulty spindle bearings, the steering 'finger' has pressed onto the spindle.

Fit for scrap: a fatigue fracture has caused hairline cracks in the gear teeth.

Also fit for scrap: this part has been over-adjusted, with the result that the steering 'finger' has pressed down on the base of the spindle.

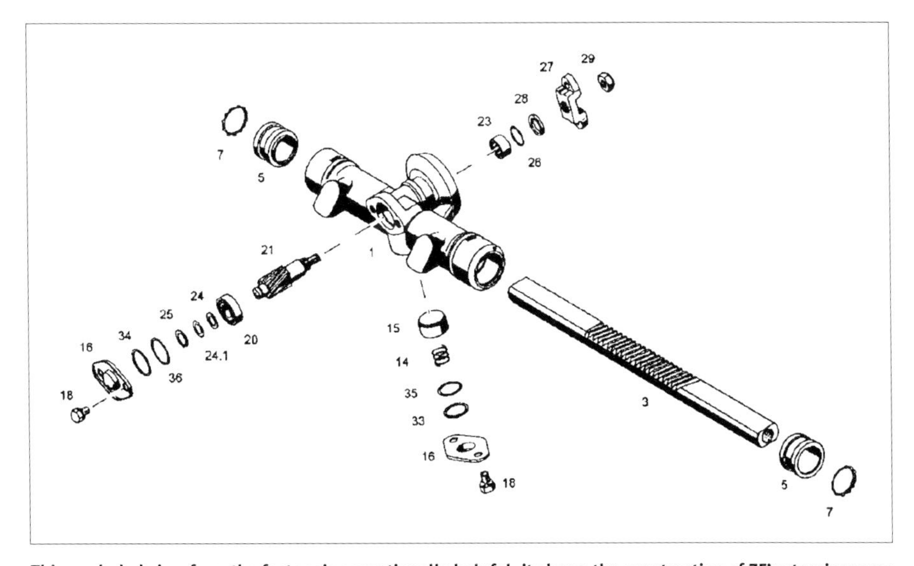

This exploded view from the factory is exceptionally helpful: it shows the construction of ZF's steering gear for the Porsche 911 and 914.

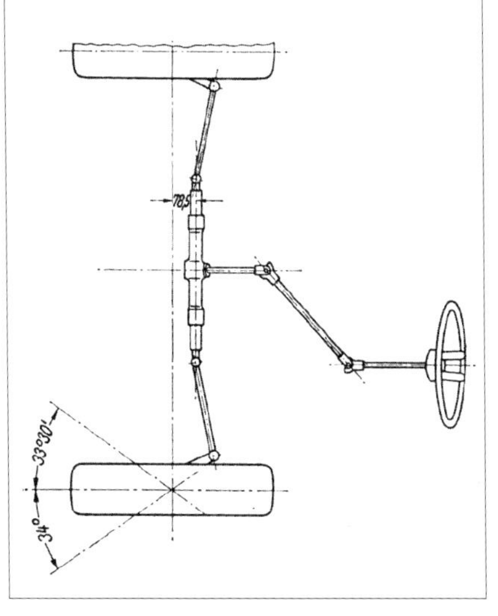

The steering system as installed in the Porsche 911: the steering column is jointed in two places.

near Dortmund as they overhauled the rack and pinion steering used in the Porsche 911 and 914. Rack and pinion systems, which transmit the steering forces very directly, were initially fitted only to extremely light vehicles, for the most part sports cars, since their gearing ratios are limited.

A rack and pinion setup transmits shocks from the road, however, more or less undamped back to the steering wheel. The steering gear consists of a pinion (usually helically cut), that meshes with a toothed rack. As the steering moves, the rack moves to the left or right, and so acts directly via the steering rods on the wheel carriers. Most rack and pinion systems are self-adjusting: a spring-loaded thrust piece (made from plastic or bronze) presses the rack against the pinion, thus completely eliminating any play. The weak points of this type of steering are the boots and bellows where they connect to the steering rods. If grease escapes and dust gets in, the rack guides wear very quickly. The only solution then is to replace the rack or the complete assembly.

The same thing applies to rack and pinion systems

If you come across an old ZF part number, you still have a good chance of finding replacement parts.

Thanks to ZF's computer network, its staff can call up information for many steering systems.

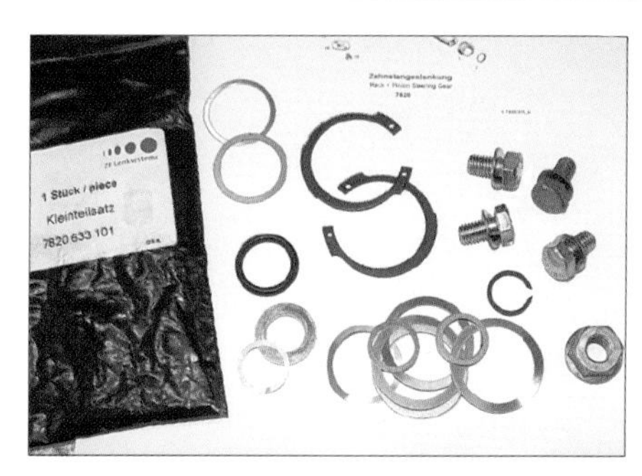

Replacement sets of seals are available for many common models of steering gear.

Preparing to reinstall the parts: before dismantling them, mark the centre position.

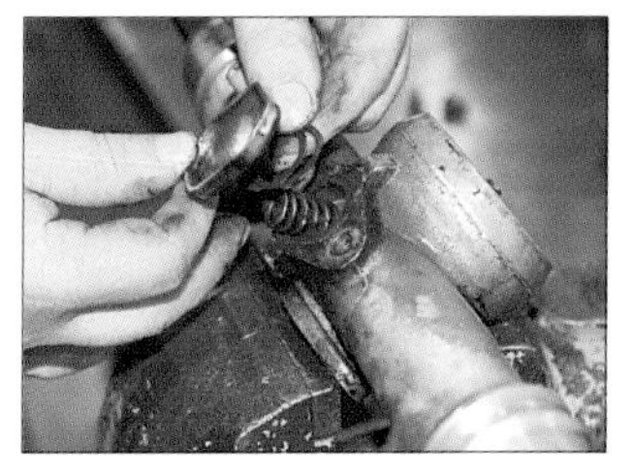

There are spacers under the pressure plate and above the thrust piece.

A spring pushes the thrust piece onto the steering rack.

The pinion is gently loosened with a rubber mallet ...

... and then removed and its condition assessed.

Now the steering rack can be taken out.

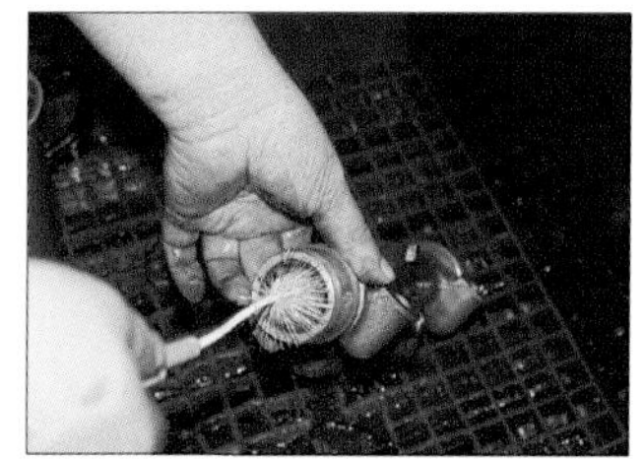

After cleaning, a visual check of the casing can be carried out.

An expert examination is also essential for steering racks.

Marks are most often found in the straight-ahead position.

These grooves on the thrust piece and steering rod are harmless.

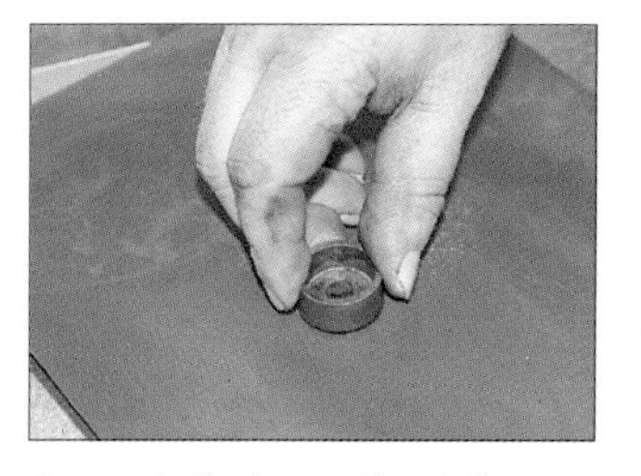

Grooves in the bronze thrust piece are ground down.

Uneven patches on the steering rack can also be dealt with.

The measuring plate shows whether the rack is straight.

The degree of curvature (which may be intended) can be found in the datasheet.

Even if they appear intact, rubber bearings should be replaced for safety's sake.

With a deft blow, they can be knocked out.

that have been damaged in an accident. In the series of illustrations shown here, the ZF technician demonstrates how to strip down and reassemble a rack and pinion system. Naturally, all the steering components must be thoroughly examined for signs of cracks, which can often occur following accidents. Damaged parts such as these should be scrapped, to avoid unpleasant surprises.

Time has also taken its toll on the old rubber bearings.

A little washing-up liquid from the kitchen helps them slip into place.

The new bearing is pressed home; a hammer may also be used.

The marking helps locate the right position to fit the pinion.

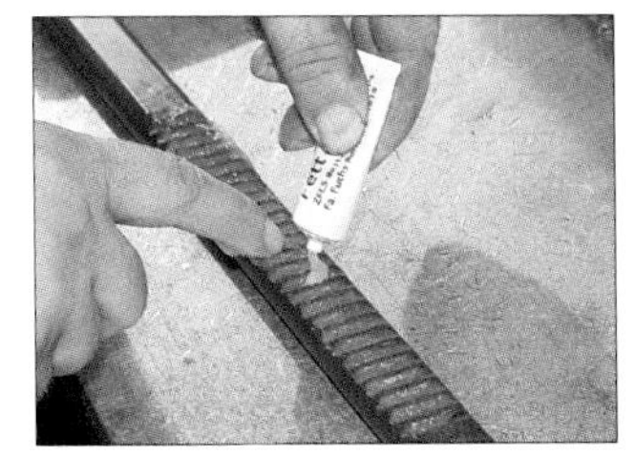

Grease instead of oil: steering racks use a specific type of lubricant.

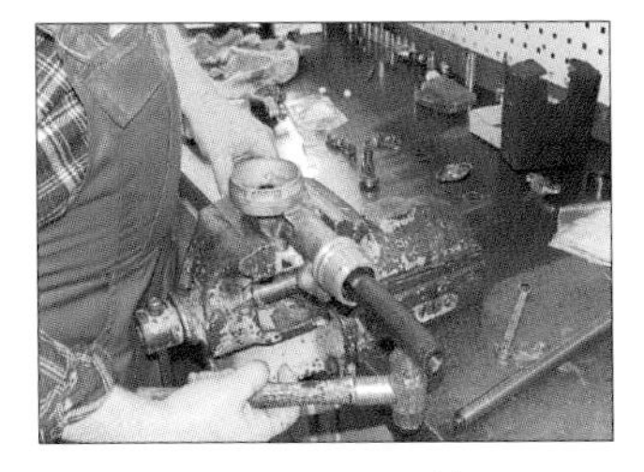

Be careful with the new rubber bearings during reassembly.

The number of spacers can be determined with an electronic gauge.

Old and new: the plate on the thrust piece (left) was distorted.

A firm seal: a special adhesive is used here.

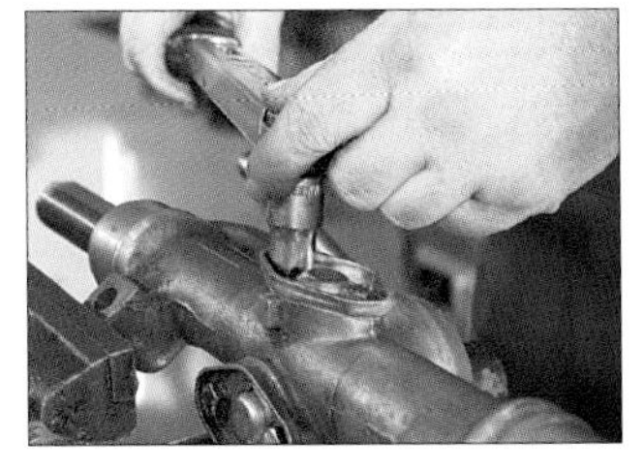

The screws are tightened using the correct torque settings.

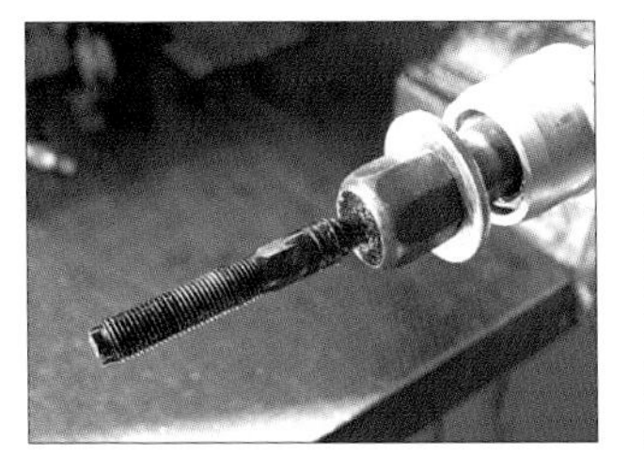

Useful during assembly: this part acts as a substitute for a steering rod.

In the foreground is the overhauled steering mechanism.

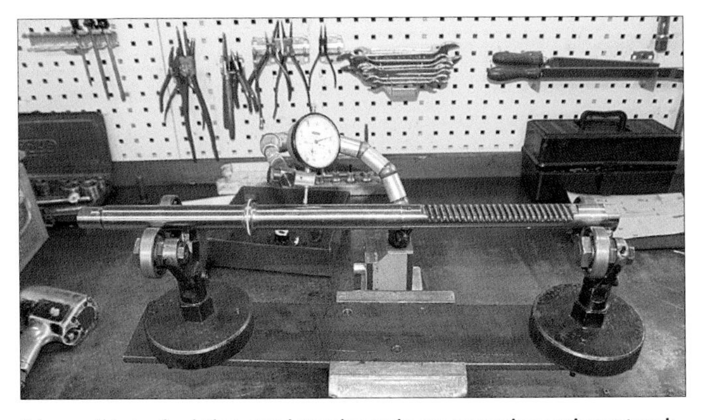

It is possible to check that round steering racks run true using equipment such as this.

The tolerance allowed can be checked by the ZF technician on the corresponding datasheet.

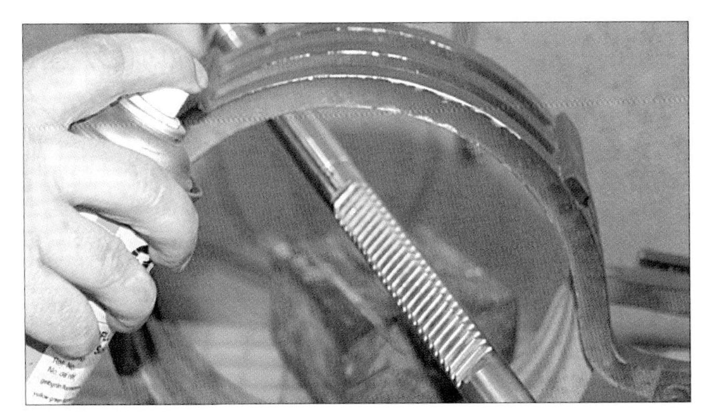

The steering rod is sprayed with magnetic powder and exposed to a powerful magnetic field.

Any cracks in the metal will then become apparent under ultraviolet light.

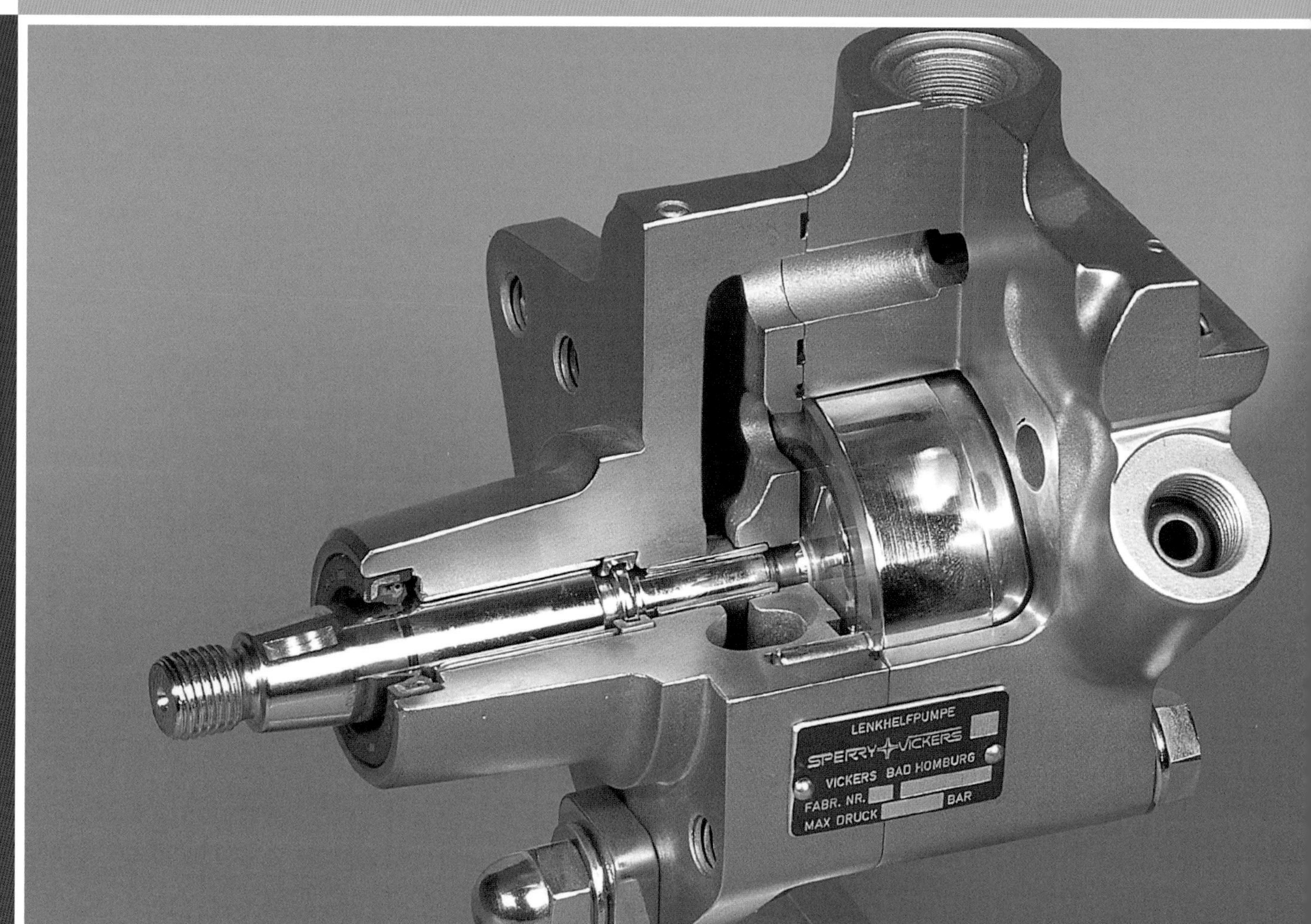
LENKHELFPUMPE
SPERRY VICKERS
VICKERS BAD HOMBURG
FABR. NR.
MAX DRUCK
BAR

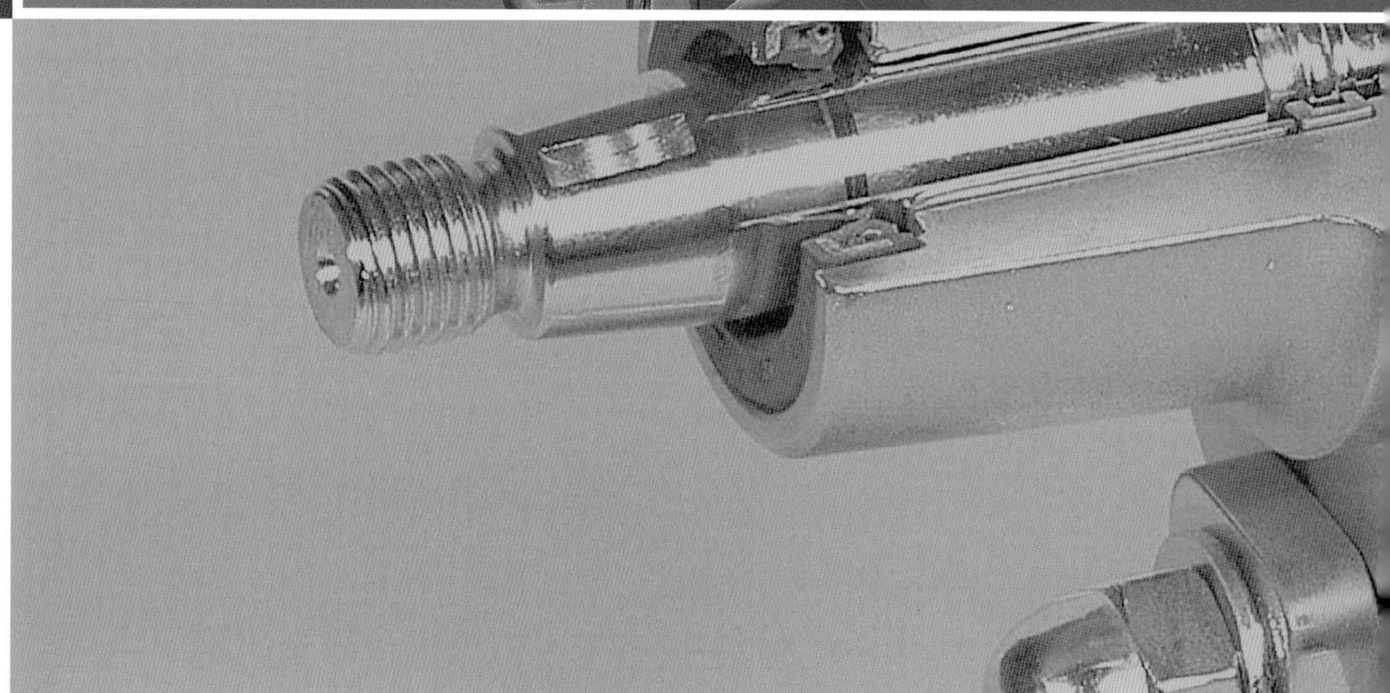

Steering
Repairing hydraulic power steering

Eventually, even the strongest drivers found unassisted steering too much: as cars became heavier, so too did the effort required to steer them. Help was at hand in the United States, though, in the form of power-assisted steering.

In 1921, Harry Vickers invented the vane-type pump, paving the way for the first power-assisted steering. The system used a pressure of at least 950psi, and was driven by a V-belt. It was another four years before the first passenger car – an Oldsmobile – was equipped with power steering, driven by a Vickers vane-type pump.

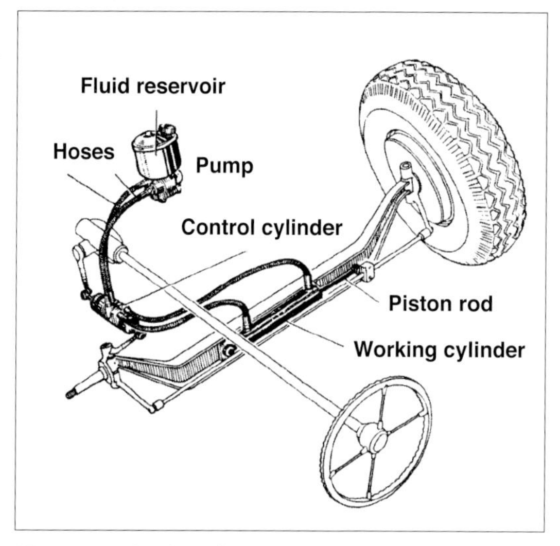

ATE power-assisted steering.

The basic design of this early power steering system is more or less the same as that of the ATE system (illustrated here), which was offered on German cars in the 1960s. This hydraulic system comprises a pump, a control cylinder mounted on the steering column, and a working cylinder installed parallel to the tie rod. Depending on the steering direction, the control cylinder directs the pressure to one side or other of the working cylinder, which then provides assistance in steering the wheels. Even if the system fails, the mechanical linkage between all the components is maintained. When the steering is at rest or at the end of a steering movement, the pressurised fluid flows via the return line to the reservoir, and then back to the pump.

The hydraulic power steering on the Mercedes W108, W110 and W111 series covered in this chapter is very similar. In this case, however, the hydraulic pressure acts directly on two working pistons inside the steering mechanism. The system is regulated by a so-called 'control rail,' which engages with the groove in the control piston. This directs the pressure to the respective working piston. At the time the pump was manufactured by Vickers Germany, a company which later merged into the vehicle hydraulics division of the LuK group. The housing for the steering mechanism was produced by Georg Fischer (GF) on behalf of Mercedes-Benz. As well as Vickers, ZF in Friedrichshafen also offered hydraulic pumps. General Motors mainly installed American Saginaw pumps, whereas English cars mainly used roller-type pumps made by Hobourn. The latter used rollers in place of vanes, and are, therefore, somewhat less efficient, but are otherwise similar in construction.

Leaks are common in power steering systems, and can lead to reduced effectiveness or complete failure of the system. The other components of a power steering system are broadly similar to those of a mechanical set-up, repairs to which we have already described.

Even if repairs are mainly a matter of replacing seals, sealing rings and O-rings, the steering system is important for the safety of your car, so only experienced mechanics should attempt to work on it. Of course, you can save some money by doing the job yourself, but, if in doubt, you should always turn to a professional!

Let's begin with the hydraulic pump. Problems with the pump will usually result in the steering feeling heavy. Before examining the pump, though, first check and, if necessary, top up the fluid level in the circuit, in accordance with the owner's manual. In addition, make sure that the belt pulley is securely located, and that the belt is correctly tensioned. Some systems have a gib which prevents the pulley from slipping on the shaft, but this can sometimes break. Inside the fluid reservoir is a filter, which should be changed every 50,000-60,000 miles (80,000-100,000km). Home mechanics often neglect this when servicing their cars, with the result that blockages are not uncommon. The gasket under the cover (often paper on older cars), may also be damaged. If none of these faults can be seen, and fluid isn't leaking from the pump housing or the steering hoses, it's likely the so-called 'rotational group' has suffered damage.

This group consists of the rotor (with slits cut into it, and vanes which expel the fluid through centrifugal force), the contour ring (which somewhat resembles the trochoidal rotor housing of a Wankel engine), and the two side plates. If signs of wear or grooves can be seen on any of these, the only solution is to replace the entire unit, which will cost ●x350 (the parts are no longer officially available from Mercedes-Benz). A two-part seal kit is also available costing about ●x220. By comparison, a new pump from Mercedes costs ●x890! If you have an old part to exchange, you can save about ●x130. It is well worth contacting the

The shape of the contour ring in the rotational group is reminiscent of a Wankel engine. The centrifugal force pushes the rotor vanes to the outer edges and the fluid flows under pressure through the side plates.

The filter in the fluid reservoir provides protection against dirt.

The pump from an old S-Class Mercedes serves as a demonstration.

The belt pulley is removed. A gib holds it in place.

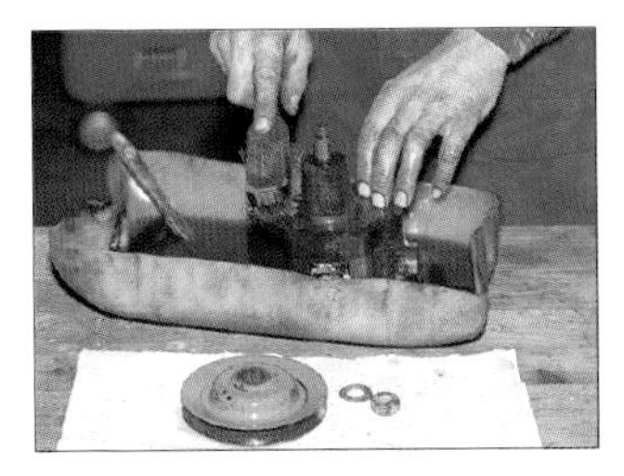
Before dismantling the housing, we clean off the accumulated dirt.

Four bolts hold the housing together from the rear.

Now the pump unit can be pulled out from the back.

The shaft is enclosed within a sealing ring and a dust protection ring.

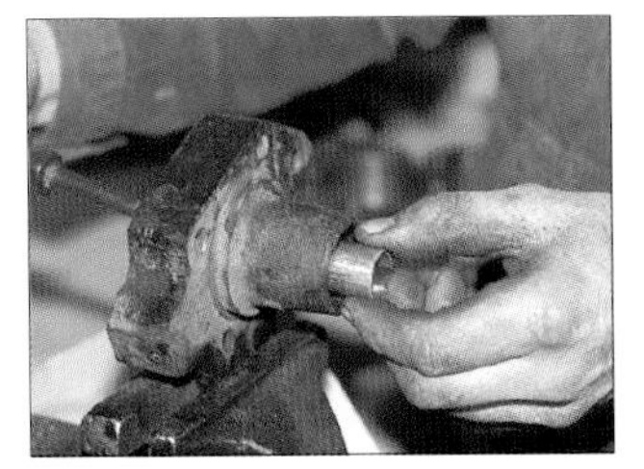
After the shaft has been removed, the bearing can be pressed out.

Is repairing it worthwhile? Often a secondhand pump will be cheaper.

official Mercedes-Benz Used Parts Centre (http://mercedes-parts-centre.co.uk).

To take things in turn, before you dismantle it, be sure to thoroughly clean the outer surface of the pump: particles of dirt are the most common cause of pumps malfunctioning. After removing the belt pulley, the bolts on the front of the housing can be undone, and the complete front part of the rotational group removed. Take care over the vanes and the position in which they are fitted. They have one straight and one rounded edge. The rounded part runs on the surface of the contour ring. The rotor is located on a serrated mount on the shaft, and can be removed.

At the front of the shaft there's first a dust protection seal and a sealing ring. We gently prised them off using a screwdriver. Often, the sealing ring has worn a groove in the shaft, and is, therefore, included as a replacement part in the

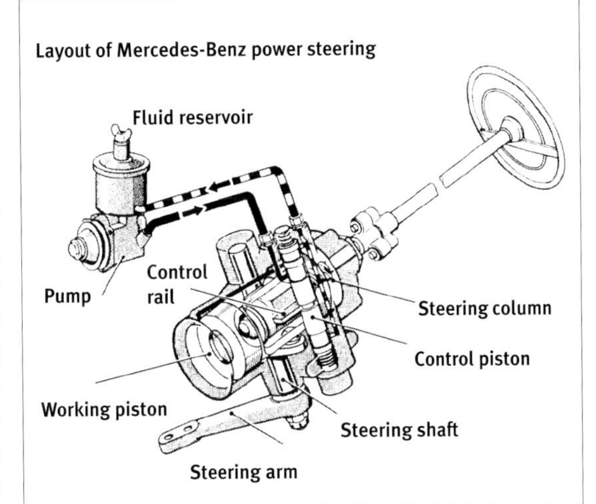

It can also be done without a press: a suitable brass bushing is helpful.

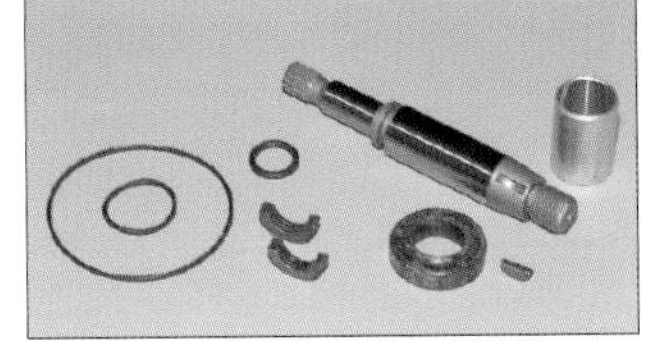
The two-part seal kit, which is available for about •x220 from Mercedes-Benz, includes a new pump driveshaft complete with bearings.

The new shaft, together with the two bearing shells, is inserted into the housing.

Next come the friction bearing bushes, which have been prepared to size.

The sealing ring is carefully pulled down onto the shaft and ...

... inserted into the pump housing with the appropriate socket.

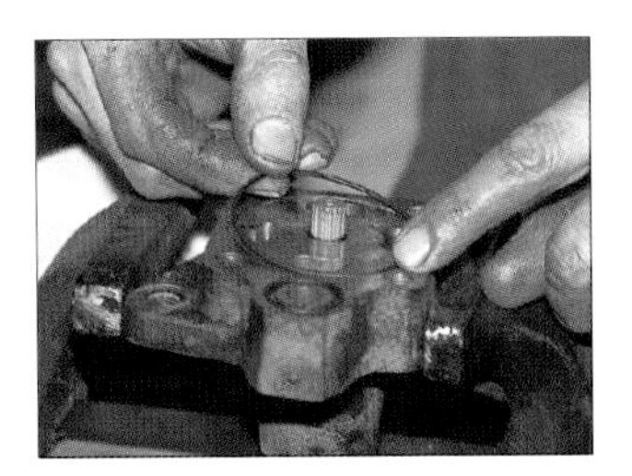
The new O-rings are fitted in the recesses provided.

The vane-type rotor sits on a serrated mount on the end of the shaft.

Only now can the ten vanes be inserted into the slits in the rotor.

Then comes the contour ring with the rearmost side plate.

This spring gets rid of the play between the vane-type rotor and the side plates.

The thrust plate is fitted in such a way that the spring sits in the middle recess.

Now both halves of the housing can be put together again.

seal kit. The axial location of the shaft is taken care of by two serrated bearing shells, which are pressed into the housing, well to the back. In order to remove the shaft, it must be pushed forward with the help of gentle hammer blows. Both the bearing shells will break in the process. Next comes a long friction bearing, which can only now be removed. If this is stuck, applying some heat may help, or – as in our case – carefully cutting into the bushing. The pump unit is now completely stripped down. Depending on the extent of the damage, either a new shaft will be fitted, together with the bearing shells and friction bearing, or a new rotational group will be used. Before reassembly, all components should be thoroughly cleaned. To press-fit the friction bearing, a brass bushing turned on the lathe to the right size should be used. An hydraulic press is ideal, but some well-judged hammer blows will also do the job.

The sealing ring and dust protection seal follow, together with the O-ring for the housing. Before fitting them, we cleaned all the parts with the appropriate hydraulic fluid. After bolting it back together, the repair to the pump unit was complete. At the back, often mounted crosswise inside the pump housing, there is another important component: the volumetric flow control unit. This ensures that at low engine speed – when parking, for example – the steering supplies sufficient pressure, while the pressure is reduced at high revs. Otherwise, you would lose all the steering feel on the motorway or freeway.

After pushing out a dowel pin from the housing, the control unit can be extracted. This runs in a finely-honed bore, which should be inspected for dirt and wear. If damage is seen on the running surface, or if the control unit only moves with difficulty, repairing it will be impossible and the entire pump will have to be replaced. We limited our work to replacing the O-rings on the metal cap under the dowel pin. A coat of black paint (be sure to mask the hydraulic connections), and the outside of the pump looked quite presentable.

If the exploded drawing of the steering mechanism hasn't put you off, you're ready to tackle the seals. The seal kit available from Mercedes costs •x130. If required, the manufacturer will also supply the complete steering mechanism as a new part (no exchange is possible) for •x4000. For this cost alone, on many cars an overhaul is the only economic option. As a rule, recirculating ball steering systems present no mechanical problems, and last virtually forever, unless they receive a sudden increase in loading (being driven over an obstacle). Now and then, problems can arise with faulty seals, which in turn cause fluid leaks or reduce the effectiveness of the power assistance. But take heart, the most important component of all, the recirculating ball mechanism, need not be taken apart to work on the seals.

After cleaning the housing, we removed the steering arm: a job for which a good puller is essential. Then the bearing cap at the side could be opened. Inside this, the steering arm shaft is guided into a needle-roller bearing. If the bearing is faulty, a replacement from Mercedes costs •x85. On the opposite side, under domed nuts, are the screws to adjust play in the steering shaft. We undid the inner adjustment screw and removed the thrust bearing; the shaft could now be removed on the other side of the housing.

Next to the adjustment mechanism and under another cover is the control piston/slide. It must be removed and examined to check that it is operating smoothly. On our car, there was some dirt inside the bore, which we cleaned using a honing tool for brake cylinders. If serious damage to the bore is found, the only solution is to look for a used steering mechanism in good condition, as the housing is no longer available on its own. Such damage is, however, very rare.

Driving it out: the dowel pin secures the cover for the volumetric flow control unit.

Don't forget: the O-ring on the cover will also be changed later on.

If the control unit looks clean and moves smoothly, it will just be cleaned.

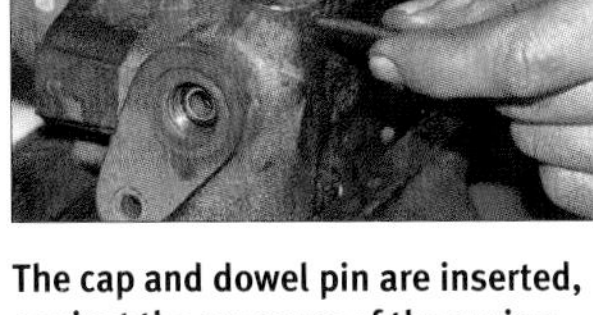

The cap and dowel pin are inserted, against the pressure of the spring.

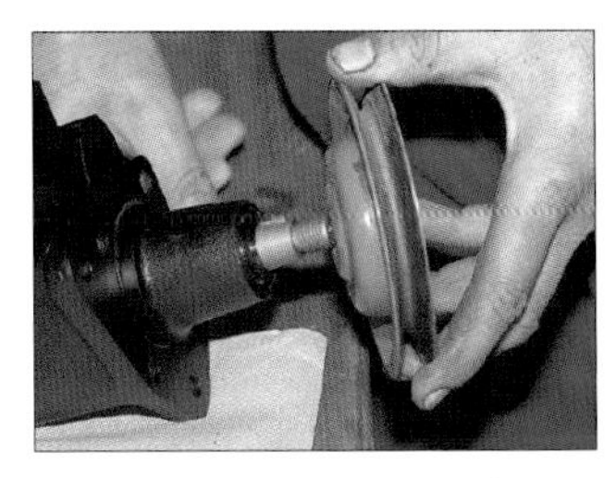

The belt pulley is bolted onto the freshly painted pump.

An overhaul is worthwhile: a new steering mechanism costs a massive •x4000.

The side cover on the steering shaft can now be removed.

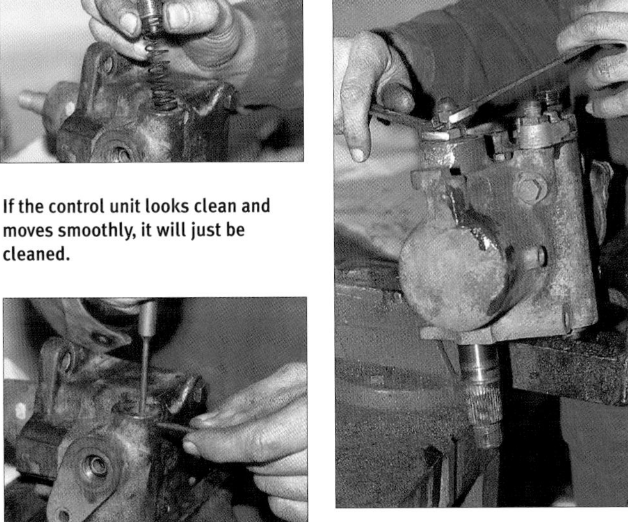

The adjustment screw for the steering shaft is under a domed nut.

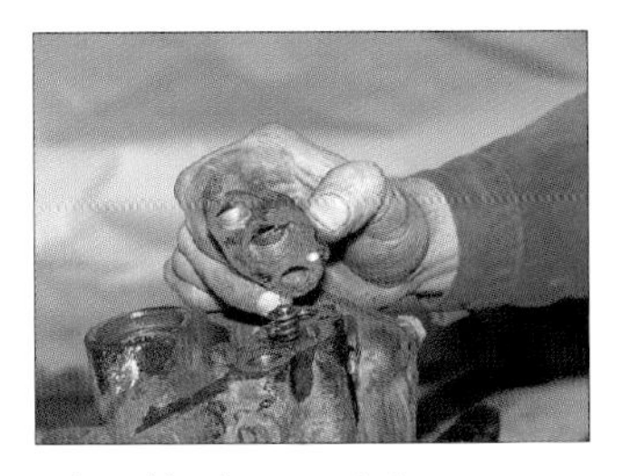

Below this, the control piston goes into a bore.

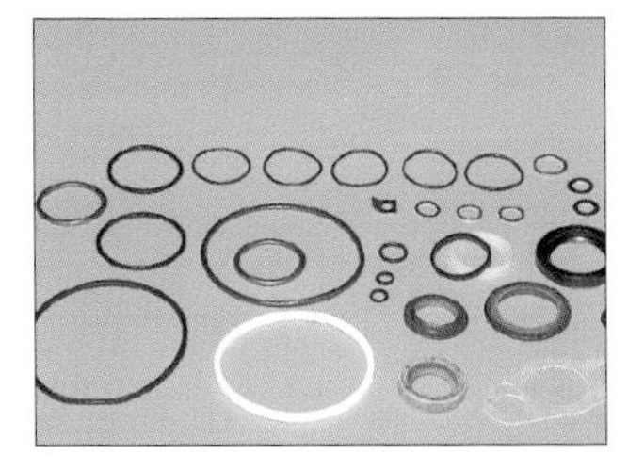

We're off! This comprehensive seal kit costs about •x220.

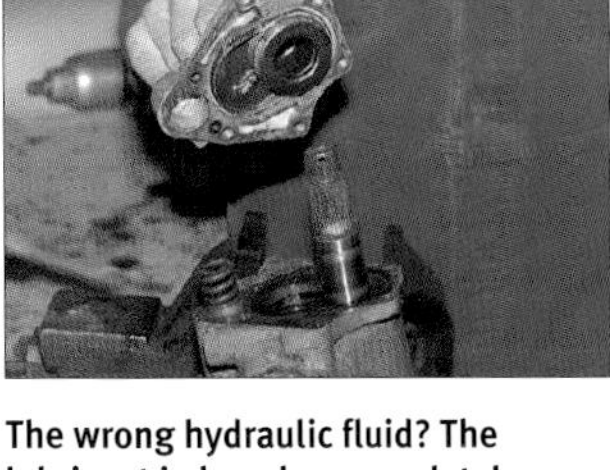

The wrong hydraulic fluid? The lubricant in here has completely gummed up.

The adjustment screw is turned all the way out. Then ...

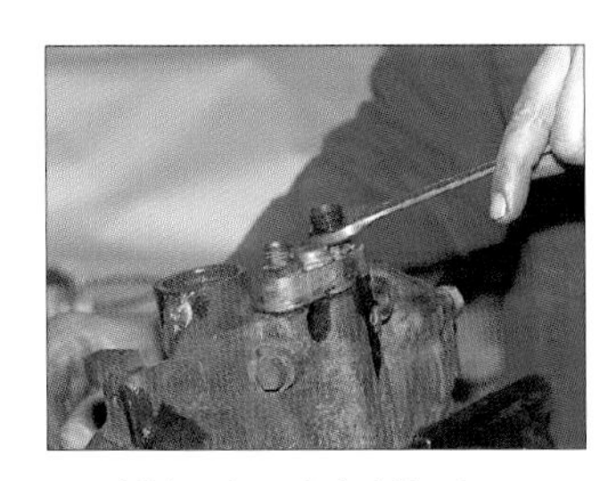

An additional cap is held in place with two hose connectors.

The piston can be pulled out of the housing. Under the final cover ...

The steering arm can be removed using a sturdy puller.

... the cap which serves as a thrust bearing can be removed, together with the steering shaft.

... is the recirculating ball unit, which can be removed in its entirety.

Finally, we removed the bearing cap on the side next to the steering shaft, and pulled out the complete recirculating ball unit. Depending on the steering direction, this part slides backwards or forwards, moving a steering shaft located in a socket. The complete steering unit works inside the housing like a piston, which moves to and fro under the hydraulic pressure. The control slide directs the pressure to the correct side according to the steering angle.

The steering unit and control rail are not rigidly connected, but rather by means of a catch spring.

The control rail engages in this way with the piston and pushes it in the appropriate direction.

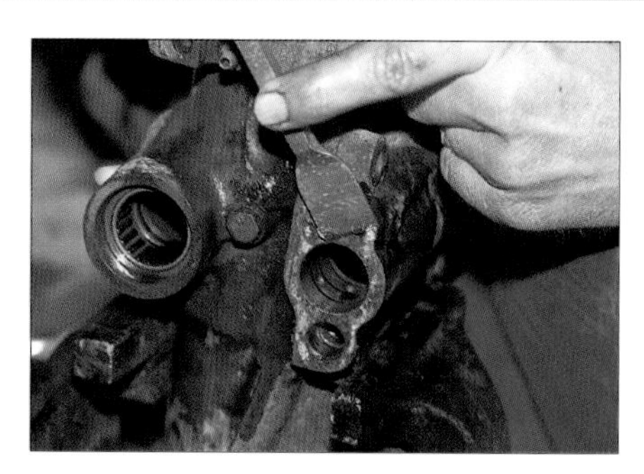
All the sealing faces on the sturdy, die-cast housing must be perfectly clean.

This circlip holds the seals tight inside the housing cover.

The position of the washers, seals and O-rings is very important.

A dust protection seal and sealing ring prevent any leaks.

Both rings are replaced. Underneath there are two roller bearings.

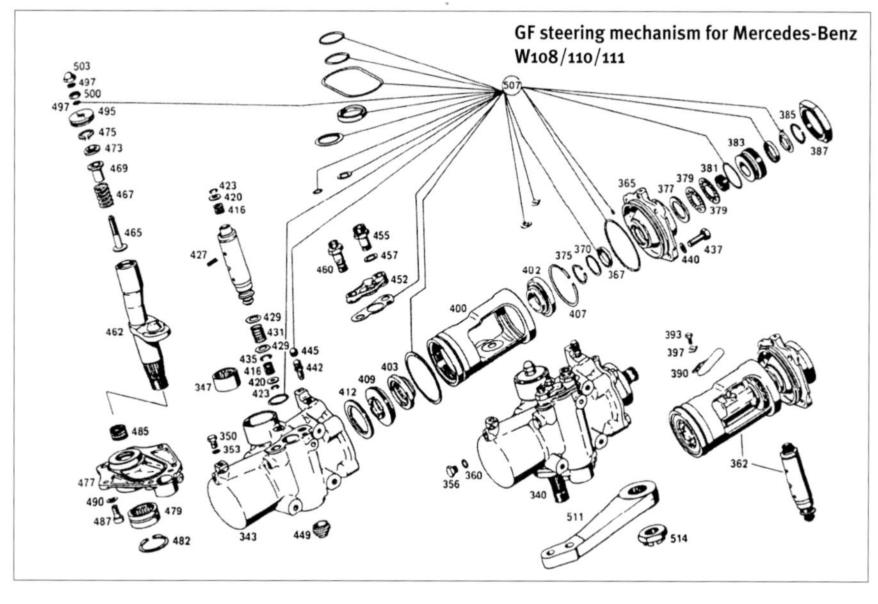

GF steering mechanism for Mercedes-Benz W108/110/111

So that the pressure can build up, at the front there is a Teflon ring, and at the back an O-ring; both are included in the seal kit. At this stage, we thoroughly cleaned all the parts. The Teflon ring could only be pressed onto the steering assembly with the help of a heat gun; the O-ring was easier to fit. Inside the housing covers there were shaft seals and O-rings, all of which were replaced.

Some O-rings are deep inside the bores, and the control slide may pass through up to five sealing rings, of which only two are readily accessible. A sharpened wooden rod can help when changing these parts.

The causes of the poorly functioning power assistance became clear as the steering gear was taken apart: on the one hand, the control slide was stiff, and on the other, the hydraulic fluid had gummed up. This suggested that a completely wrong type of hydraulic fluid had been used in the past. We cleaned all the parts with brake cleaner, and rubbed a little fresh hydraulic fluid onto them before reassembly.

Positioning the control rail correctly in the groove of the control slide is tricky. If all goes well, you can insert the recirculating ball unit and turn the control rail – which is lightly held in place – onto the side on which the control slide will be inserted. The control slide should then be pushed in until it is about 2cm (0.8in) short of its end position. Insert the control rail into the groove by turning the steering shaft, and, at the same time, slowly push the control slide home. If it has worked, the control slide should move easily to and fro as you turn the steering from side to side.

Reassembling the remaining components is the reverse of taking them apart, using the new seals from the repair kit.

The manual states that play in the steering shaft can only be adjusted using specialised measuring equipment, but there is another way. With the steering mechanism removed from the car, the play is roughly adjusted so that all the parts move freely. With the mechanism back in the car, turn the steering from side to side a few times with the engine running, then switch off the engine. With one hand on the universal joint for the tie rod and the other on the adjustment screw, it's easy to feel when there's no more play in the universal joint. Crude though this may sound, this point can be accurately determined if you have a good feel for it, and indeed many garages adjust the steering in exactly this way.

An O-ring and a Teflon ring are fitted on the steering unit.

Take care: one of these small seals can quickly be overlooked.

Inside the bore of the control piston there are five O-rings altogether!

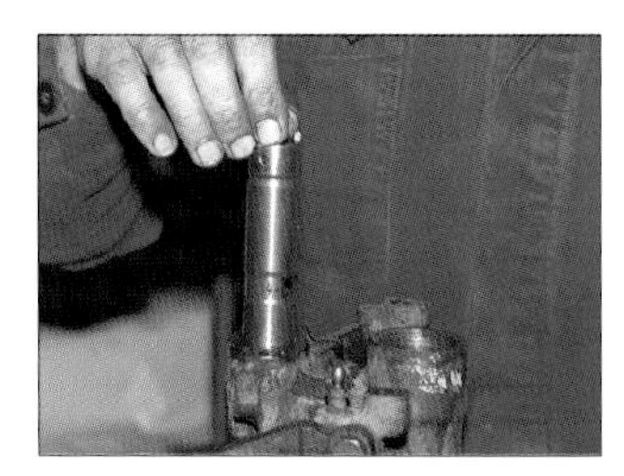
The control piston is inserted using plenty of lubricant.

All the gaskets are replaced as the unit is reassembled.

The hose connectors are fitted using copper sealing rings.

Now the steering shaft returns to its operating position.

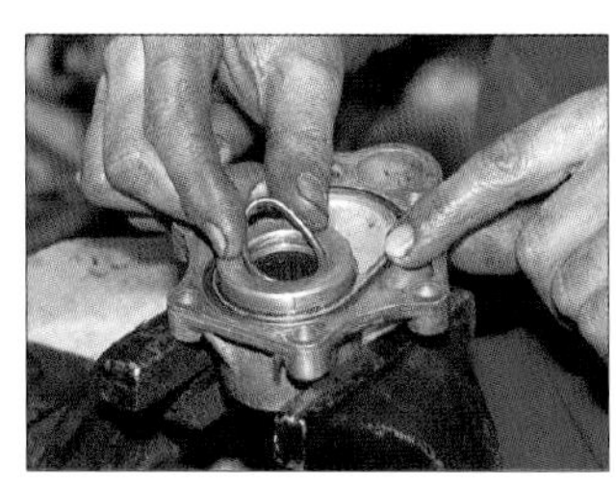
Inside the steering shaft cover is a thick O-ring.

On the outside, too, there is a dust protection seal and a sealing ring.

Even this needle-roller bearing comes at a price: it will set you back •x85.

The play on the steering shaft is only roughly adjusted to begin with.

The cornet screws used when adjusting the front suspension.

Some models have provision to increase the pressure when the engine is idling during low-speed manoeuvres, using the connection here.

A cotter pin reinforces the linkage between the crown nut and the steering arm.

Later, the hydraulic system is bled with the engine running.

You should reckon on one to two days to overhaul the servo pump and steering gear – and you'll save a good deal of money by doing it.

The servo pump for the Mercedes W108/110/111 models is available from stock. It costs about •x900. Used parts from cars that have been scrapped are often very cheap.

Overhauling the power steering pump and steering gear only makes sense if the hoses are still in good shape. These can sometimes be expensive, but construction machinery repair workshops often manufacture their own high-pressure hoses, and may be able to help you out.

With the servo pump running, open the bleed screw until fluid comes out without bubbles, then check the fluid level again. Older fluid reservoirs often have a paper gasket inside the cap: these should be replaced each time the reservoir is opened, as they frequently lose their shape and let in dirt and moisture. If you follow the schedule to replace the filter, you should enjoy your newly overhauled steering for a long time to come.

STUURBEKRACHTIGING

Steering
Retrofitting electric power steering

Fitting modern electronic power steering to a classic – sacrilege or salvation? We looked at the conversions offered by a Dutch firm, and allowed ourselves to be persuaded. Nowadays, many classics are already fitted with modern power-assisted steering.

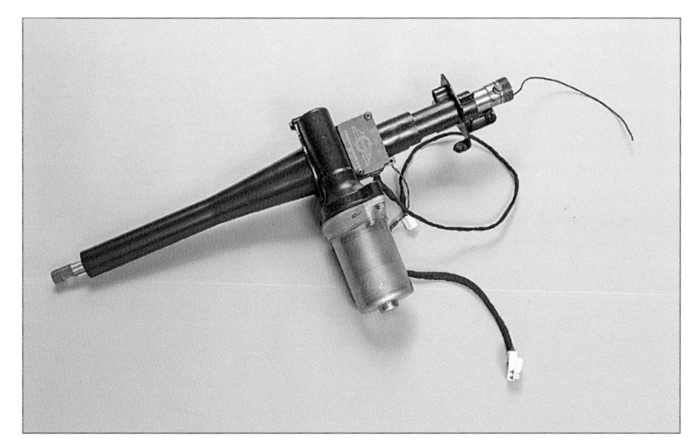

A compact steering aid: only the steering column has to be changed.

Not a big deal: Roger Reijngoud (on the right) explains EZ's modern power steering; in the background, the Jaguar Mark 2 waits for its system to be fitted.

Even heavy cars from the 1950s and '60s can be converted.

Off-the-shelf parts: the electronic power steering systems which EZ fits are originally mass-produced for a variety of new cars.

The blessings of modern car electronics will not, however, necessarily suit all classic car enthusiasts. On a car like an 'Adenauer' Mercedes from the 1950s, who needs ABS, ESP, rain sensors or a headlamp warning buzzer? In classic cars it comes down to the genuine, unalloyed pleasure of driving, with all the charm of mechanical systems rather than electronics. But sometimes you will work up a sweat as you grapple with the heavy steering on a tight bend, wishing that it was a bit more comfortable to use. And even the best of spouses may consistently refuse to take the wheel of your 'lorry,' which is just too heavy to drive!

This is where modern technology can provide an elegant remedy. The Dutch firm EZ Electric Power Steering offers electric power steering conversions for classic cars. For the 'Adenauer' Mercedes, everything is available as

If the original steering columns have to be adapted, EZ has plenty of welded sleeves to hand.

The different stages in making a steering column bracket: the finished product is at the top.

The steering columns (split into two sections) are made from solid metal: the blanks will fit nearly every car.

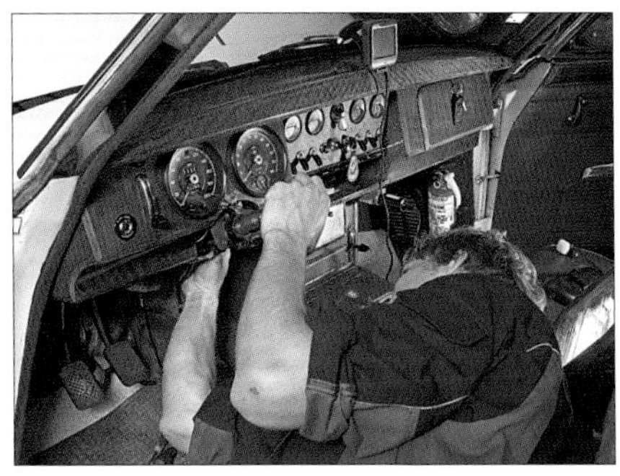

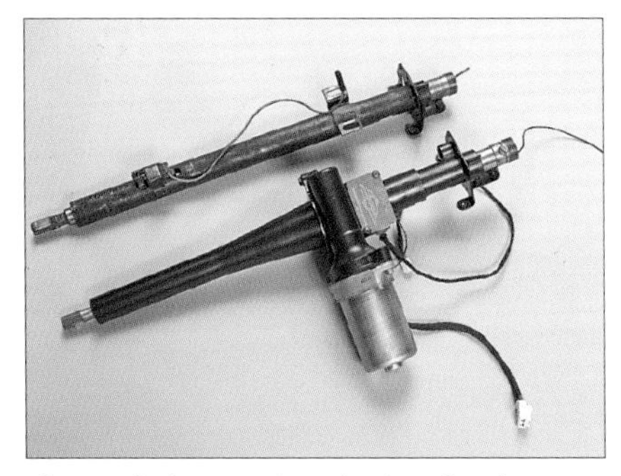

With the bonnet open, there is scarcely any difference to be seen in the Jaguar's engine bay after the conversion. All the key changes take place out of sight, and that goes for the interior as well. First, the original steering column is removed; in the picture on the right, it can be seen above the preassembled EZ conversion kit.

a preassembled kit – and our test drive in a 300 Coupé, with its light and precise steering, was a real pleasure!

Electronic power steering is neither a backstreet garage bodge nor a radical innovation. "For our conversions, we use thoroughly developed parts, which have been tried and tested in mass production," explains Roger Reijngoud, one of EZ's directors. "On new cars in high-volume production, this space-saving technology is well established and works reliably. The particular challenges we face, however, are in adapting it to the steering column, and accommodating the modern components in cars which were never designed for them."

The electric steering system can be explained very quickly. The steering column is in two sections: where they are split, a torsion bar with a sensor placed on it bridges the two parts. When the torsion bar twists as the steering turns, the sensor sends an electronic signal to the control unit. The control unit is, in turn, connected to an electric motor, which provides assistance to the steering. The degree of assistance is, therefore, dependent on the load placed on the steering, and the speed of the car, since the extent of torsion of the bar is directly transmitted by the 'black box' and the sensor to the electric motor, the speed of which is adjusted accordingly. When driving straight ahead, no signal is sent by the sensor on the torsion bar, and the power assistance remains inactive.

Roger Reijngoud explains the advantage of the load-sensitive function: "In hydraulic power steering systems, the pumps are driven by the motor in all operating conditions. This additional equipment naturally draws some power: as much as 4bhp may be lost. Furthermore, the servo pump uses hydraulic fluid, and leaks can therefore occur. And, by no means least, there is a drive belt which can wear with age or break. If the hydraulic power steering fails, there is always a safety risk, as the car will become considerably heavier to steer. When electronic power steering has been fitted later, this doesn't matter."

This safety margin should not be dismissed: apart from the steering column, no other part of the drivetrain

In situ: EZ's steering is bolted in exactly the same position as the original equipment.

Installing the wiring, including that for the steering-column controls, is the most time-consuming job.

A perfect fit and no gaps to be seen: the electrical wiring and servo are well concealed.

To the right: as you can see, you can't see anything! Monsieur Dutruel and his son are most impressed: after six hours' work, their 1967 Jaguar Mark 2 is ready for its power-assisted drive home to France.

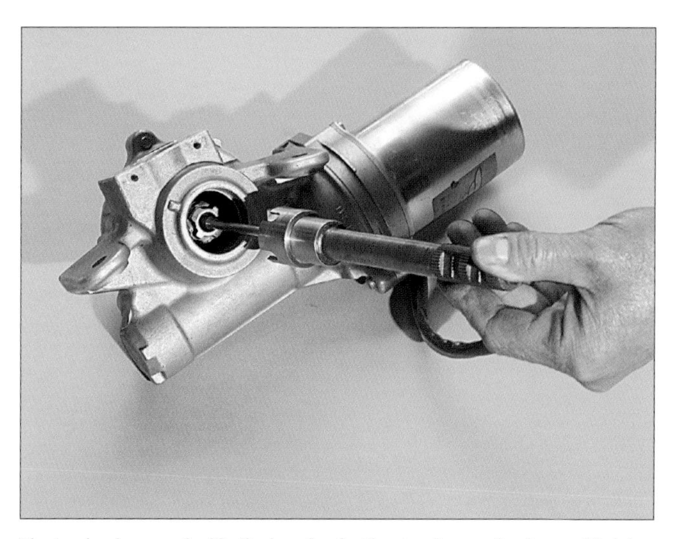

The torsion bar runs inside the housing for the steering mechanism and twists in response to the steering force applied.

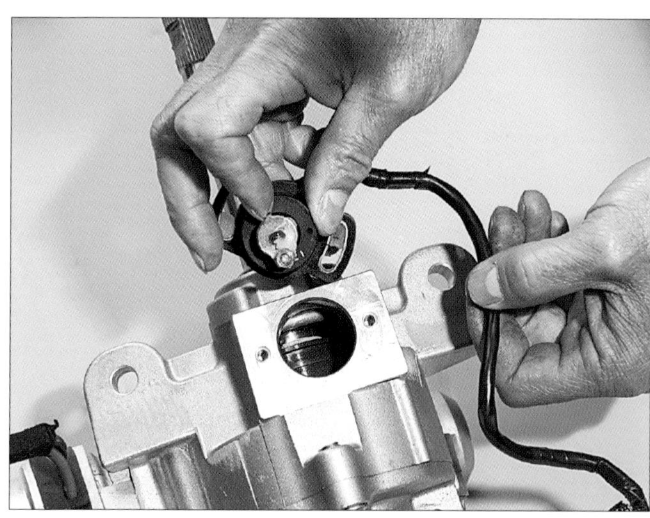

This sensor measures the extent of torsion of the bar and transmits this information as an electronic signal to the black box.

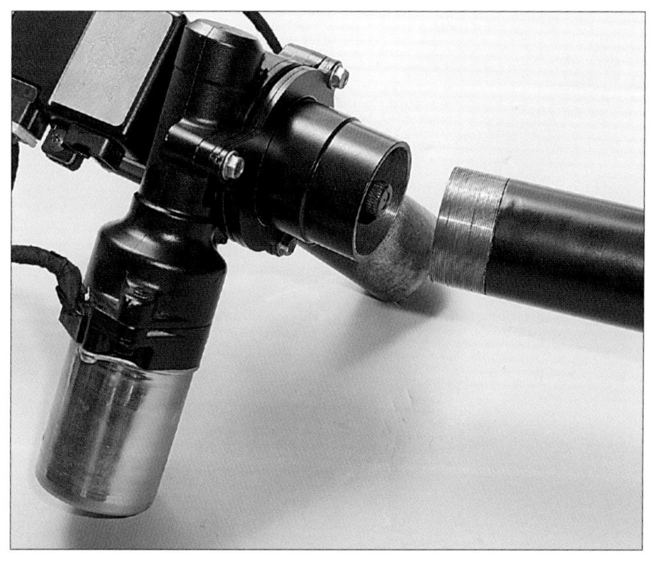

In this case, the original casing tube for the Jaguar Mark 2's steering column can be bolted in place.

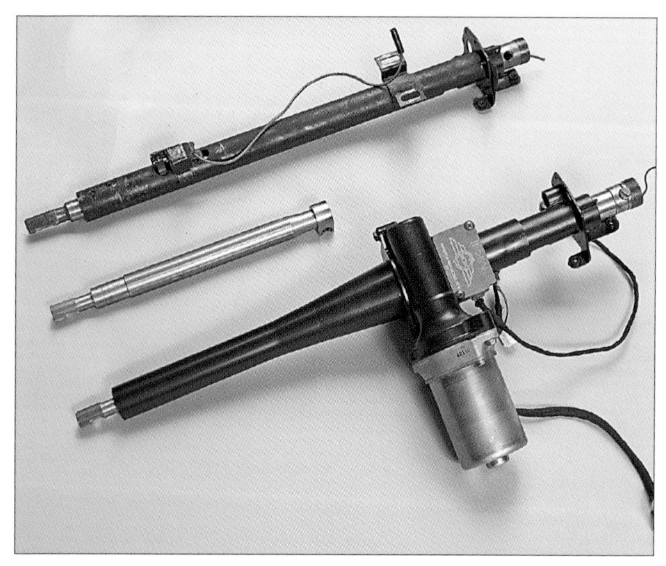

The original, a blank and the new part: the Jag's steering column was easy to fit – a stroke of luck.

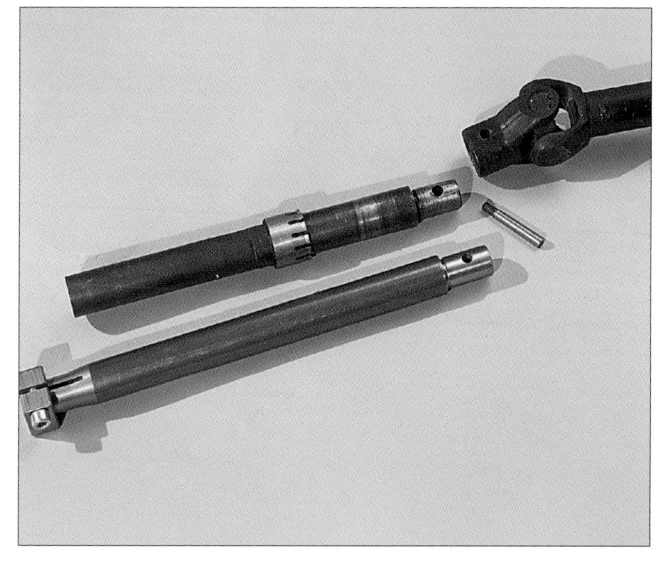

Anything goes – even a steering column to the customer's desired length, as long as there is enough room in the car.

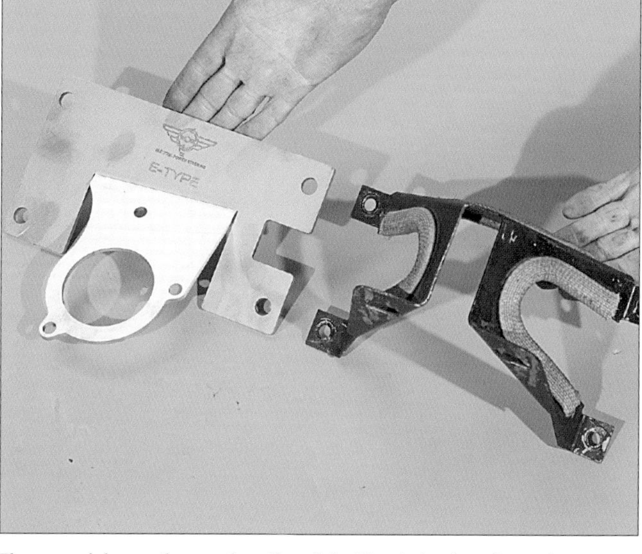

These special mountings replace the original brackets – here from a Jaguar E-Type.

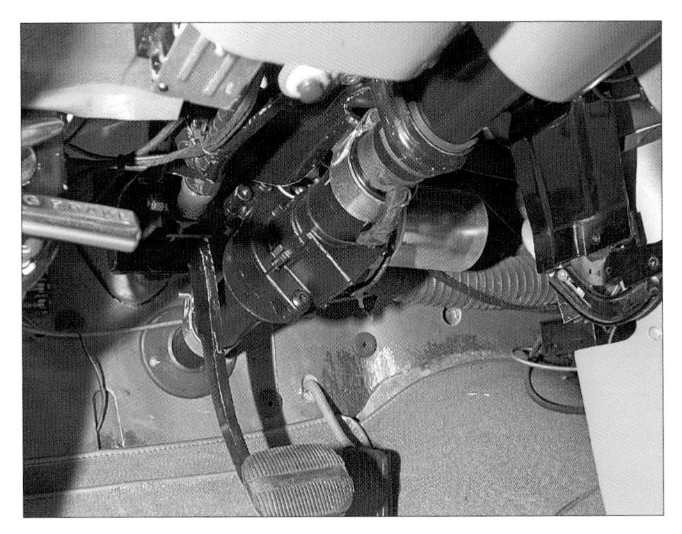

In the Chevrolet Corvette C1, the power steering unit is almost out of sight. It is continuously variable from light to firm, to suit the driver's taste.

is altered, so that, should the system fail, the steering will simply continue to function as it did before it was converted. And, since EZ's conversion is controlled electronically, the degree of assistance can be adjusted by means of a rotary knob on the dashboard, which controls a potentiometer added to the circuit. If so desired, it can simply be switched off altogether: it only needs a rocker switch to interrupt the power supply!

More work required: on the Mercedes 300 Coupé, the steering-column gearchange also had to be re-routed.

No need to worry about a loss of power: the servo motor draws very little current, and its electrical power consumption could only be detected when the engine was not running.

The work undertaken on the original structure of the car is straightforward, and, above all, the conversion is, in general, easily reversible. Only in exceptional cases is it necessary to use the original parts from the car. To quote Roger Reijngoud: "When installing our systems, for the most part we fabricate our own steering columns, incorporating the appropriate connectors at each end for the steering wheel and steering mechanism. The diameter and strength of the metal used for car steering columns are very similar, so it is exceptional for us to have to split an original part and fit the mountings for the torsion bar.

The conversion is complete; the test drive to adjust the settings was successful.

Almost like a new car: the Mercedes' steering responds to the author's delicate commands.

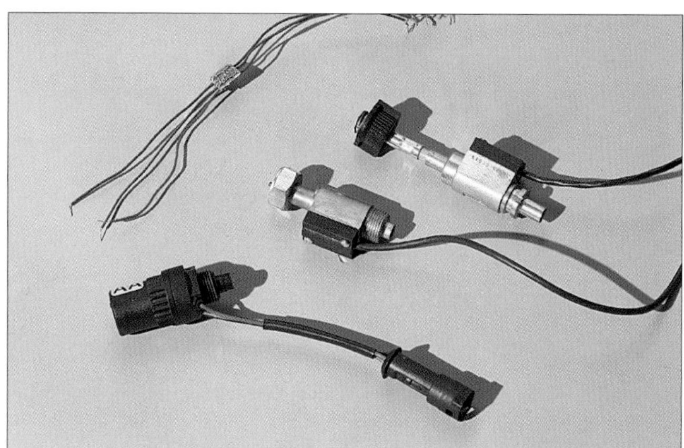

Depending on the configuration, the speedometer also sends a signal to the black box for the steering. On new cars, the speedometer drive is electronic; EZ Power Steering uses special adapters.

Left: the devil is in the detail. A slip ring provides the contact for the horn.

It is more often the case that we need to fit outer sleeves or coverings; fortunately, this is rare on the majority of classic cars. The trickiest job really is to modify the electrical fittings around the steering wheel.

"In the case of horns especially, the column and its outer shroud are often part of the electrical circuit. By using dual sliding contacts, one behind and one in front of the servo drive connected to the steering column, we have managed until now to deal with every type of circuit. Even the power connection for the steering itself isn't rocket science: it just needs positive and negative connections – all the other cables run to the little black box, which houses a processor that handles the signals. If there is enough room, we can attach the black box directly to the steering; if not, we will look for an unobtrusive location under the dashboard."

And unobtrusive EZ's power steering certainly is. The steering column on most cars runs under the dashboard, where there is usually enough space to accommodate the parts without affecting the driver's legroom. On cars where the steering column is covered by a shroud, absolutely nothing can be seen after the conversion has been carried out, and, in any case, no modifications are needed in the engine bay. The conversion of a Jaguar Mark 2 which we followed in EZ's workshop in the Dutch village of Herwijnen, provided a good example of this. At first glance, the only before-and-after difference was the much more comfortable steering. The servo, complete with its specially produced mounting brackets, was completely hidden. Fitting took just a few hours, as a complete steering kit is available off the shelf for the Mark 2. "Every model of car which we convert for the first time goes through our workshop, where the process is fully documented. When the first car is finished, the conversion kit is produced at the same time. In the case of popular cars such as the Jaguar or the Mercedes 190 SL, it's worth our while to produce several steering kits straightaway. By now, we have more than 50 models of car in our range. And it is no longer necessary to have steering kits which are already available fitted by us in Holland. We have authorised workshops throughout Europe. Naturally, even with standard kits, we can cater to special requests from our customers. It is no problem to make steering columns longer or shorter, to create a more comfortable driving position. Drivers who until now have fought shy of fitting a smaller sports steering wheel, because they feared the heavier steering this would cause, can go ahead now without any worries – the power steering will compensate for the reduced leverage of the smaller wheel. Even on cars fitted as standard with an adjustable steering column, we always find a way of keeping this extra feature." The same applies to steering-column gearchanges, such as that fitted to the 300 'Adenauer' Mercedes that we drove on test. The linkage for the steering-column lever had to be modified to fit around the servo motor. In this case, there was no alternative, but the modified gearchange is reliable and properly set up. It all means more work though.

The fear that suspension components such as the tie rod ends, kingpin bushes or steering mechanism may be damaged by the new power-assisted set-up is unfounded: the forces exerted on the operating parts from the steering wheel are no greater than before. They will just be transmitted more evenly. Nor will the power consumption of the steering overload the dynamo. With the engine running, the ammeter of a converted Chevrolet Corvette C1 failed to register any deflection, even when turning the steering wheel rapidly at a standstill. With the engine turned off and the ignition on, the steering drew just five amps – too little to do any serious damage to a battery in good condition during parking manoeuvres.

Talking about damage, the power steering systems

The next stage in development: the electric motor drives two gear wheels. On the right is the working model, on the left the almost finished prototype.

An Aston Martin DB2 can now also be driven by self-confessed softies. If a real man comes along, you simply turn off the power assistance!

themselves – mainly manufactured by Mitsubishi – meet the high safety standards normally required of new cars. As we have already mentioned, if the electronics fail on an old car, nothing will happen, as the original steering will go on working as normal.

When the power assistance is out of action, it has no effect on the steering and represents no danger whatsoever. Even if a mechanical fault arises on the steering column, the steering wheel will not spin freely all of a sudden. That could only happen if the torsion bar were to break. In this case, however, the ends of the two steering column sections would immediately mesh together: they have a system of crown-like teeth, which provide a join if the torsion bar shears off, so that the damaged car can at least be parked.

The electronics are also dependable; until now, EZ Power Steering has had only one complaint. "A customer from Estonia reported that the steering on his car had failed after a few thousand miles. We had the car shipped to us and set about finding the fault. The reason was simple: an oxidised circuit board inside the black box. When it rained, water could get in, and after a short time it crippled the part. With a new control unit, which we located in a different position, the problem was solved," Roger Reijngoud told us. "The supply of replacement parts for our steering systems is guaranteed: we deliver to each of our authorised workshops, in case anything should go wrong."

Naturally, some questions remain, which may be critical in purchasing decisions for many customers. How much does it cost? What cars is it worth fitting to? And, above all, have the roadworthiness testing authorities in your country given EZ Power Steering's conversions their blessing, especially for cars with historic registration status?

The price for such a conversion, including installation, ranges on average from ●x2150 to ●x2500, but this will vary according to the type of car. The Jaguar Mark 2 was converted for ●x2150, but for more complicated designs such as the 'Adenauer' Mercedes, as much as ●x3400 may be due. Even that amount may be fully justified in relation to the value of the car: it would be difficult to find one of the rare original ZF power steering systems for the Mercedes any cheaper.

It is usually not worth modifying small or light cars, as the steering is light to use, even without any power assistance. A Fiat 500 doesn't need power steering. It is certainly possible and might be a real option, if the car were to be converted for use by a disabled driver, for example. In rallying too, the switchable power assistance can also come in handy. As Reijngoud says, "Nowadays, we have conversion kits for relatively light cars such as the Saab 96 or BMW 02 Series, which are often used in motorsport. The customer is king: we can fulfil every wish. We require only that the work be carried out in our workshop or by one of our partners, as these are safety-related parts, and it is essential that they are properly fitted."

As far as the German safety inspection and historic vehicle registration procedure is concerned, there are currently two sides to the coin. Since 2009, EZ Power Steering's kits have received ISO 9001 certification from the TÜV in northern Germany; a test engineer only needs to certify that the kit has been fitted correctly. The costs are similar to those for a registration entry. The situation regarding historic vehicle registration status is more complicated. Classic cars to which such equipment has been retrofitted no longer correspond to their original specification; nor can electronic power steering be considered a period accessory. This even applies if the car in question was originally available with hydraulic power steering, since that is a different system technically. A conscientious vehicle examiner should, therefore, withdraw the converted vehicle's historic registration status or refuse to grant it in the first place. However, it is always worth checking directly with your local testing station, preferably before having the conversion carried out. Sometimes, vehicle examiners have more leeway than you might expect, especially as the installation is reversible, contributes to driving safety and, in most cases, is fitted completely unobtrusively, with the result that the visible condition of the car remains unchanged.

MZ Z571H

Tyres
Are wide tyres the best option?

"Only a wide tyre is any good." It would be easy to come to this conclusion when summarising the developments of the past 30 years. But is the widest piece of rubber the best choice for classic cars? We checked it out.

First of all, we need to distinguish between two things: the actual width of the tyre's contact patch, and the tyre's aspect ratio; that is, the relationship of the sidewall height to the tread width. Wide tyres have existed for a long time, but modern-day wide tyres are primarily defined in terms of their height, which has continually come down in relation to their width. In 1962, a tyre with an aspect ratio of 88% was considered a low-profile tyre! Three years later, it was superseded by a 'Super-Low-Profile' aspect ratio of 80%. In 1968 even lower profile tyres followed, this time known simply as '70 Series.' This naming convention has continued until the present day: the width and height of the tyre are shown on the sidewall. With 185/70-sized tyres, the sidewall height represents 70% of the tread width. Nowadays, we have reached aspect ratios as low as 30%.

Three main factors account for this trend to fit ever wider and lower profile tyres. First, car wheels continue to grow in size, to make room for ever larger brake discs and more complex suspension componentry. Secondly, cars are constantly gaining in weight. Since the volume of air inside a tyre determines to a large extent the load it can carry, its width has to increase as its height decreases, to avoid reducing its load rating. Thirdly, wide tyres demonstrate tangible benefits in handling and in controlling high acceleration and braking forces.

Dancing in the rain: on the slippery steering pad, the skinny tyres are in their element.

They can be held in a steady drift on the wet surface.

In the dry, our search for the handling limit was accompanied by an almighty screech.

With wider tyres, the car sat lower on the dry tarmac. The lateral acceleration, however, was only marginally higher than with the narrower tyres.

When the steering pad had been sprayed with water, the 185-section tyres performed almost as well as the narrower 165s.

The intermediate tyres in our comparative test remained almost as easy to control on the slippery surface as the 'skinnies'.

Too much is too much. The 185-section tyres were somewhat harder to catch in a drift than the narrower standard tyres.

When it came to the Fiat 124 Sport that we used as a test car, we logically had to accept that two of these three points were not valid: brake and suspension components do not automatically increase in size by replacing the standard 13in wheels with 15in rims. The unladen weight of our Fiat 124 Sport (980kg/2160lb) also remained virtually unchanged.

We investigated the changes to the car's handling at the driving safety centre at the Nürburgring. Michelin's competition department provided a rapid pit stop service, so that we could quickly swap the three different sizes of tyre we tested – ultimately, the ambient temperature also affects the grip that tyres provide. With the help of these experts, we were able to conduct our tests in virtually identical conditions – in as far as that is ever possible at the Ring …

The procedure was always the same: a quick warm-up run along the twisting country roads near the circuit, then onto the steering pad at the driving safety centre, where the maximum cornering speed could be measured with an optical sensor, first on a dry surface and then wet. Finally, we determined the effect of the changes to the rolling resistance on the car's top speed.

We tested tyres in three different widths, which were equipped with tyres in three different widths, which were fitted to various types of light-alloy wheel. First, the 5Jx13 Cromodora wheels which were available for the Fiat when it was new, fitted with the standard 165HR13 tyre size, known today as 165/80 R13 83T. The second combination was also available back in 1971: 185/70HR13 (available today as 185/70 R 13 86T) tyres, mounted on 5.5Jx13 Abarth CD 30 wheels from Cromodora – used then as now by Fiat drivers with sporting pretensions. The third combination was a true product of the eighties: 7Jx15 Gotti wheels with 195/60 R15 tyres.

We began with the 165-section tyres, which, by today's standards, seemed strangely lost inside the Fiat's wheelarches. As we had driven to the Ring on the

Period-correct appearance: the 124 Sport was presented with these wheels in the 1969 brochure.

A whiff of Monte Carlo: the works competition cars ran on Cromodora CD 30 alloy wheels.

Eighties-style: the low-profile tyres fitted to 15in Gotti rims look very modern.

A lot of tyre, not much wheel: 80-Series rubber was the norm in the 1960s.

Due to their slightly lower profile, the 185/70 tyres are almost as tall as the 165/80s.

A lot of wheel, not much tyre: the fancy 15in wheels are conspicuous in the foreground.

A look inside the wheelarch shows how far out the wheel and tyre come.

The reduced offset of the CD 30 is the decisive factor here. It broadens the track …

… and modifies the steering offset. The Gotti rims are offset further inwards.

185/70 tyres, we immediately felt some wander when turning into corners or taking S-bends on the twisty country road. Under the influence of the centrifugal force produced when cornering, the tall sidewalls allowed the alloy wheel to shift to and fro above the tread area, until a firm pressure point could be established. Even driven enthusiastically, the car could then be steered around the corner with a fair degree of accuracy.

As the cornering speed increased, though, a heartrending wail could be heard from the skinny tyres, which led us to doubt whether they were really a match for our sporting ambitions.

On the dry steering pad, the narrow tyres struck up their sad refrain once again, at speeds as low as 20mph (32kph). It took a real effort to increase the speed nonetheless. At 32mph (51kph), we finally got there: the Fiat's live rear end gently broke away – not least because the inside rear wheel could no longer put its power down on the road, and the outer wheel – which was now bearing the car's entire rear weight – was overloaded. Despite the tyre squeal, however, the car could be controlled without difficulty with a single hand on the steering wheel.

Next on the agenda was the inner part of the steering pad, which had been sprayed with water and was as slippery as a layer of snow. Here, the skinny tyres were in their element. The car kept to its line in a good-natured way, and any attempts to break away could easily be corrected. The wretched screeching sound was finally over, too. At this speed, we were miles away from experiencing any aquaplaning. The theory also confirmed the optimum results achieved by the narrow tyres.

The braking test proved challenging. Without ABS, every driver needs a certain amount of time to get a feel for the amount of grip from his or her tyres when braking. On a dry surface, everything worked out fine during our test; on wet tarmac, however, the results were all over the place, making a reliable comparison impossible. On dry tarmac, a stop from 31mph (50kph) gave a stopping distance of 48.6ft (14.8m). On the autobahn, the standard tyres earned extra points. At 115mph (185kph) in fifth gear, the rev counter needle reached the redline – that ought to do it ...

Time to change the tyres. The CD 30 wheels with their scorpion logo were reminiscent of the works set-up on the 124 Abarth Rallye, but the 185/70 R13 tyres seemed very conservative by today's standards. Even when manoeuvring in the car park, the increased steering effort was apparent. Over the next few miles on country roads, the extra 2cm (0.8in) tread width was noticeable. The car felt firmer and sportier, and there was much less tyre squeal when cornering fast. The maximum possible cornering speeds seemed to have increased dramatically; the car felt more confident and predictable during changes in direction in S-bends.

On the steering pad, the tyres began to squeal much later, at 24mph (39kph). At 34mph (54kph), things gradually came to an end: again, the outer rear wheel broke away in a controlled manner once the grip had been lost on the inside. Despite the remarkably modest increase in speed, the car was appreciably safer and more predictable before it broke away. It remained forgiving, too, as it started to drift. On the wet steering pad, the difference compared with the narrower 165-section rubber was so slight that it was impossible to say straightaway which size was better. The 185-section tyres broke away more suddenly, but could not make up for this shortcoming with improved handling. It was a different story when braking on dry tarmac: the wider tyres scored better, with a shorter stopping distance of 46.6ft (14.2m).

The return run to the final pit stop was also along the autobahn. On the same section as before, the engine would only pull reluctantly up to the redline. The increased rolling resistance was perceptible, while the frontal area of the car had also increased, thanks to the wider tyres, which also marginally affected its aerodynamic resistance. Even with a long run-up, the game was over at just above 112mph (180kph).

Michelin's mechanics were ready with the Gotti wheels, although these were not officially approved for the Fiat. Oh well, we wanted to give them a quick try anyway. Manoeuvring the car in the car park, the increased effort needed at the steering wheel was apparent. Cars fitted with wheels such as these generally have power steering nowadays, but not the 1971 Fiat. As soon as the car was under way, though, the suspension felt more modern, firmer and stiffer. The tyres transmitted the smallest bumps in the road back to the steering. At the same time, you had the feeling of driving a finely-honed piece of machinery, which – provided you concentrated on it all the time – could be placed with complete precision. The turn-in to corners was an absolute delight, and not a peep was to be heard from the 195-section tyres. The car took corners quickly and unspectacularly, if you overlook the point that the tyres tended to veer slightly to the right when going over small bumps in the road. This made driving quickly great fun, although the typical feel of driving a classic was lost.

On the dry steering pad, things really picked up. The wide tyres followed their line without protesting. It needed a smaller amount of turn at the wheel to follow the set radius. A little squeal set in at 28mph (45kph), but the car would zip around at about 31mph (50kph). With regard to its handling, we were expecting a real increase in the cornering speed, when – whoa! – the tail suddenly spun round and could not be caught. We tried again, gradually upping the speed to the limit. The car felt safe until just before the end, and then at 34mph (55kph) the Fiat broke away once more. Drifting the car was a balancing act, these tyres did not like half measures. This impression was confirmed on the wet steering pad. First, as usual the front wheels of the car pushed wide, then the tail broke away with virtually no warning. As we expected, it proved even more difficult to catch an immediate spin. In short: on wet surfaces, the wide tyres were really treacherous!

On a wet road the wide tyres become treacherous, and the car can break away unexpectedly.

The tail pushes out in response to the centrifugal force with even the slightest throttle application.

Aquaplaning also becomes worse as the tyre width increases.

It usually begins like this: as you turn into the corner, the car runs stubbornly straight ahead as the front wheels understeer and push wide.

In addition, the aquaplaning tendency increases in relation to the width of the tyre – an effect which can be especially dramatic with lightweight cars. When braking on dry tarmac, the wide tyres regained some ground as far as safety was concerned. Their stopping distance of 44ft (13.4m) was the shortest in the test. On a wet road, however, it was very hard to modulate the brakes with the 195-section tyres.

Above the 100mph (160kph) mark, the car struggled somewhat, and, despite a longer run-up, could only accelerate to 112mph (180kph). The bottom line: with wider tyres, the cornering speeds that could be achieved did increase, but by much less than we had expected. The difference was more clear-cut when braking: wider tyres do offer an improvement in terms of safety. The handling characteristics are a strength of wide tyres with a lower profile; sporting drivers will certainly accept the trade-off in terms of comfort. On the limit, narrower tyres are more forgiving and easier to control when they break away. The same applies on wet surfaces, and if aquaplaning occurs. On the motorway or freeway, the narrower tyres win out with their lower rolling resistance and reduced frontal area, which also helps reduce fuel consumption.

In our case, the happy medium was the best choice. The 185/70 tyres were sporty enough not to begin squealing on every bend, and remained controllable on wet surfaces. The precise handling response and direct road feel of the 195/50 tyres had too many downsides, among them increased wear on the suspension components: the tie rod ends, wishbone bushes and wheel bearings were not designed for these increased loads. Furthermore, 15in wheels with low-profile tyres are not appropriate historically on a car from the 1970s. They could be an issue when having the car inspected for registration as an historic vehicle. Even with a normal registration plate, the Gotti wheels are no longer acceptable today, as they have not been officially certified. The CD 30 wheels, on the other hand, *have* been approved. They are the ideal compromise between the optimal tyre width and the necessary contact pressure. Individual preferences for a sportier or more comfortable ride play their part in this decision. If you don't have the opportunity to investigate this as thoroughly as we did, it's worth talking to fellow club members or leafing through old tyre tests.

Tyres

Tyres for classic cars: market overview

These days, the availability of tyres for classic cars is pretty good, with the exception of extremely rare exotica. But even then, classic tyre specialists can often find the right solution.

A classic car naturally needs the right tyres. A Borgward Isabella with low-profile tyres and the wheels to match would be like your grandma wearing a pair of brightly-coloured platform shoes.

For sure, grandmothers today like to look hip, but when it comes to old cars, it's not always easy to respond to the demand for tyres of a type last current 20, 30 or 40 years ago. The tyre stockist on the nearest trading estate will most likely shrug his shoulders and say: "We don't carry those, we can't get them any more." His laconic response, usually accompanied by a stern expression, makes it clear that there is nothing more to be said. Fortunately, though, classic car enthusiasts have their own suppliers. A quick online search or a flick throuhgh some classic car magazines will return a lengthy list of classic tyre suppliers, such as Longstone Tyres (longstonetyres.co.uk), Vintage Tyres (vintagetyres.com), The Blockley Tyre Company (blockleytyre.com), and Coker Tires in the US (cokertire.com).

Robert William Thomson (1822-73) invented the pneumatic tyre, but was then forgotten.

Well-known brands such as Firestone, BFGoodrich and others, who have long since given up the licence and manufacturing rights for their historic vehicle tyres to specialist companies, scramble for customers alongside manufacturers such as Michelin, Vredestein and Kumho, who are more or less actively involved in the market for older and more recent classics.

We will restrict this overview to tyres for classic cars. The tables on the pages that follow are already so detailed that adding a similar outline for goods vehicles or motorcycles would be beyond our scope. In the last few years, things have changed quite a lot. According to MOR in Munich, for example, Cooper Tires has put nearly all sizes of its tyres into production; Maxxis now offers an alternative (primarily supplying radial tyres for American cars). Vredestein has also stepped up its commitment, whereas Dunlop has faded into the background somewhat in Germany, despite producing a new run of SP Sport D8 tyres for the 300 SL 'Gullwing' Mercedes a few years ago.

The market for tyres to fit older and more recent classics is therefore always in a state of flux, so it is worth referring to the suppliers' websites or enquiring whether the range of tyre sizes, brands or common tread patterns has grown, or indeed been cut back. There are many companies active in this area who can supply the right tyres from multiple brands for virtually any car, regardless of when it was built. The newly remanufactured Pirelli Cinturato tyres, for example, which were a standard fit on the Lancia Aurelia and Flaminia, among many fascinating cars, are once again available.

Pneumatic tyres, which were invented by the Scotsman Robert William Thomson in 1845, and then reinvented by John Boyd Dunlop in 1888, emerged at the end of the 19th century. These pneumatic tyres were initially white and had no tread pattern. From about 1905, anti-skid strips were placed on the then common beaded-edge tyres to prevent them from sliding on slippery surfaces. Soon afterwards, grooved tyres came into fashion, and finally, Michelin added carbon to the compound: car tyres became black. The removable wheelrim, which became popular as a result of Ferenc Szisz's win for Renault in the very first Automobile Grand Prix in 1906, ultimately ensured that pneumatic tyres prevailed over the traditional solid-rubber tyres for cars and motorcycles. Later, changing the wheelrim gave way to replacing the entire wheel.

For goods vehicles, this transition took a few decades longer.

The high-pressure beaded-edge tyres were mounted on flat-base rims, and had sidewalls with beaded edges to provide a firm seal. This type of rim was popular above all with the straight-sided high-pressure tyres with wire beading common in the USA, and made fitting tyres considerably easier. This form of tyre remained the norm

Beaded-edge tyres were originally white, but became black when carbon was added.

The tire bead sits in the groove on the edge of the wheelrim.

A must for early American cars: the typical straight-sided tyres.

Michelin's Bibendum tyres come in metric sizes.

in Europe and North America until the end of the 1920s. When beaded-edge and straight-sided tyres were eventually superseded by low-pressure crossply tyres mounted on drop-centre rims (using textile cords – which were later reinforced with steel – stretched crossways in the carcass), Michelin introduced its special Bibendum tyres. The French manufacturer wanted to overcome the specification in inches which had been customary until then and delivered its tyres in metric sizes – which, to begin with at least, led to a dead end.

Valentin Schaal, the head of MOR in Germany, manufactures the most common beaded-edge and crossply tyres for vintage cars with an American business partner and distributes them under the Excelsior brand. The brand was originally used by the Hanover Rubber Works, which later became better known as Continental. MOR acquired the moulds for Continental's old crossply tyres, which in the meantime had supplied the replacement tyre market in South America, together with the rights to use the Excelsior name and revive the old brands, as also happened with Deka (tyres for military and commercial vehicles) and Phoenix (whitewall tyres for Mercedes-Benz cars from the 1970s).

Cheers! From 1904, the 'Michelin Man' was also fond of a drink.

In the early days, anti-skid strips provided a form of tread.

Schaal's company also supplies beaded-edge tyres from other brands, such as Waymaster, Firestone, Coker and Dunlop. These include white beaded-edge tyres too, which were common before the First World War. Finding the right tyres to fit to these venerable classics built before 1930 is therefore still entirely possible. The same holds good for straight-sided tyres, which are now supplied almost exclusively by BFGoodrich.

Like their historic predecessors, modern beaded-edge and crossply tyres still require tubes. The remanufactured tyres from Excelsior, BFGoodrich, Dunlop, Michelin, Firestone or Waymaster are up-to-date in terms of their production technology and use modern rubber compounds. They will last for 12,500 miles (20,000km) or more, depending, of course, on how they are used. Excelsior's crossply competition tyres with H or V speed ratings are, moreover, approved for use at speeds of up to 149mph (240kph). They have the tread pattern typical of their period, in order not to affect the car's original appearance.

Two developments gave a further impetus to the tyre market in the late 1940s: the introduction of tubeless tyres, and the market-readiness of Michelin's radial-ply 'X' tyre, with its cord plies arranged radially rather than diagonally. Avon, Dunlop, Kumho, Michelin and Vredestein can all supply suitable crossply or radial tyres (tubeless or with inner tubes) for the majority of older and more recent classics still on the road, as they appeared after the Second World War, and in the traditional sizes. The earliest versions of this new generation of tyres had a sidewall height of 85% in relation to the tread width; from the 1960s and '70s onwards, these gradually led to low-profile or super-low-profile tyres with aspect ratios of 70%, 60%, 50% or even less. Manufacturers such as BFGoodrich, Michelin, Maxxis, Bridgestone, Kumho and Pirelli offer the most common sizes, for American cars as well. Although not cheap, Michelin can also supply replacement TRX tyres, which the French developed in the seventies.

If you take your newly acquired tyres along to the nearest tyre fitter, in order to have them fitted to the appropriate wheels, however, you may often be greeted with baffled looks, at least as far as really old tyres are concerned. Beaded-edge tyres, for instance, can only be fitted by hand, which requires skill and knowhow, as well as a good deal of physical effort. Before it is fitted, the tube for such tyres must be dusted with talcum powder. The tyre must then be pulled onto the rim using a tyre iron, and finally, correctly positioned by hand. At the same time, care must be taken to ensure that the bead toe slips evenly into the beading on the inner flange. After a final check to make sure that the valve is properly seated, the tyre can be inflated with air.

An experienced specialist, who can carry out this job for you (or at least show you in detail the first time you attempt it), is not the worst choice on this occasion ...

Tyre-fitting machines are often used with crossply tyres fitted with tubes, but the fitter must always apply talc to them and make sure that the tube does not become trapped between the tyre and the wheelrim and that it is free from creases. Thoroughly cleaning rust and dirt off the rim well and, in the case of wire wheels, applying tape to the rim (to prevent the spoke mountings from wearing through the tube) are all a routine part of the job.

Beaded-edge tyres (with inner tube)

Size	Load rating	Brands available
28x3	53M	Excelsior, Firestone*
30x3	72M	Excelsior, Firestone*
30x3 111	76M	Excelsior, Firestone*
710x90	76M	Excelsior, Waymaster*
760x90	76M	Excelsior, Waymaster*
i 810x90	NA	Waymaster*
815x105	80M	Excelsior, Waymaster*
875x105	88M	Excelsior, Waymaster*
730x110	76M	Excelsior
820x120	89M	Excelsior, Waymaster*
880x120	89M	Excelsior, Waymaster*
895x135	99M	Excelsior, Waymaster*

Straight-sided tyres (with inner tube)

Size	Wheel rim	Brands available
31x4	23inch	BFGoodrich*
32x4 111	23inch	BFGoodrich*
32x4 111	23inch (load rating 92P)	Dunlop
33x5	23inch	BFGoodrich*
32x4	24inch	BFGoodrich*
33x4 1/2	24inch	BFGoodrich*
34x5	24inch	BFGoodrich*
33x4	25inch	BFGoodrich*
34x4 111	25inch	BFGoodrich*
35x5	25inch	BFGoodrich*
34x4	26inch	BFGoodrich*
36x4 1/2	27inch	BFGoodrich*
37x5	27inch	BFGoodrich*
36x4 111	28inch	BFGoodrich*

Bibendum tyres (metric)

Size	Load rating	Brands available
130/140x40	with inner tube	Michelin*
150/160x40	with inner tube	Michelin*
12x45	with inner tube	Michelin*
12/13/14x45	with inner tube (LI 80P)	Michelin
13x45	with inner tube	Michelin*
13/14x45	with inner tube (LI 88N)	Waymaster
14x45	with inner tube	Michelin*
14/15/16x45	with inner tube (LI 84P)	Firestone
15/16x45	with inner tube	Michelin*
14/15/16x50	with inner tube (LI 93P)	Dunlop
160/65R315	tubeless (LI 73S)	Dunlop

*** = Load rating not available, therefore only suitable for cars built before 1929**
NA = no data available

Fitting a tube inside a tubeless tyre without thinking is, on the contrary, a dangerous mistake, which has already caused many blow-outs: the inner surface of tubeless tyres is much coarser than that of tyres with tubes. In this case, there is a severe risk of rubbing through the tube! The only exceptions are those tyres which have been expressly approved by their manufacturers for use with inner tubes. You should find out in detail beforehand about the clearance certificates issued by the different manufacturers. With tyres such as these, the inner surface is smooth, but the main factor when determining whether a tube is needed is the wheel itself, which must be airtight.

When buying new tyres, you should pay attention not only to the correct size, but also to the load rating and speed rating. The load rating indicates the highest load a tyre can bear. We have summarised the weight in kilograms corresponding to each load rating in the table at the end of this chapter. We have done the same for the different speed ratings, which are denoted with capital letters. Whether you are looking at the load or speed rating, it is better to choose a tyre with the next higher rating rather than a borderline value. It goes without saying that the rating should never be lower than that indicated on the papers for your car. Otherwise, it may not be legal to drive, and your insurance will be invalid as well!

The same is true when switching from crossply to radial tyres. For classic car enthusiasts, such a change may be relevant if the original tyres are no longer available, or on cost grounds. In individual cases, it may be worth consulting a vehicle examiner you trust. He or she should generally be able to help you further. Some vehicle testing stations, may also be able to offer advice on tyre-related issues. The examples given in the overview on the next page, which also includes aspect ratios for modern low-profile tyres, should only be used for general guidance: they also depend, among other factors, on whether the wheel clearance remains adequate at maximum lock and with the springs fully compressed, after switching to the appropriate new tyre size. Take care: depending on the extent to which the rolling circumference has changed, the speedometer reading will no longer be correct!

Another issue concerns whitewall tyres, which many classic car owners would like to fit to their pride and joy – either because they match its original appearance or simply because they look better. Nowadays, original whitewall radial tyres are available for nearly all American cars; the difficulty, however, arises with the tyre sizes which were common in the Old World.

We can't say anything more here about the right way to choose and fit tyres, but these topics – and also the subject of winter tyres for classic cars – are explained in detail on the websites of the tyre manufacturers and suppliers.

Vulnerable whitewall tyres naturally call for particular care. It is essential to clean them regularly, so that the white does not lose its brilliance (although this is inevitable as the tyres age). Heavy dirt can be successfully removed when tackled with a scouring pad from the kitchen. The warnings about loosely fixed whitewall bands, which in the past many classic car fans simply stuck on behind the rim, in order to save a bit of money, have since been repeated often enough that they are now common knowledge. But it is worth saying here once again: stay away from these products which can cut slits into the rubber – they can wear right through the tyre sidewalls and lead to dangerous blow-outs. They are a false economy. Our comprehensive tables listing the tyres currently available ~~in Germany~~ for older and more recent classics include competition tyres which are road-legal, such as those from Avon, which has

Narrow crossply: Dunlop Chevron

1920s-style Avon Dixi

Excelsior 19in: an early 'wide' tyre

The tread pattern becomes ever more modern

Dunlop's Sport crossplies ...

... are available in several versions

Dunlop Forth 4HW for tough conditions

CROSSPLY TYRES

Size	Load rating/ inner tube	Brands available	Size	Load rating/ inner tube	Brands available	Size	Load rating/ inner tube	Brands available
4.80-10	58P/TT	BFGoodrich	7.10-15	95P: 95S / TL	BFGoodrich	4.75/5.00-18	80P; 75P / TT	Goodrich; Excelsior
4.40-12	56P / TL	BFGoodrich	7.10-15	99P/TT	Firestone	4.75/5.25-18	NA / TT	Michelin
4.50-12	56P / TL	BFGoodrich	7.60-15	99P / TL	BFGoodrich	5.25/5.50-18	84N; 88P; 90P/TT	Waymaster; Dunlop
5.20-12	68P / TL	BFGoodrich	7.60-15	99S / TL	BFGoodrich	5.25/5.50-18	84N; 88P; 90P/TT	Excelsior
5.60/6.00-12	75P / TL	BFGoodrich	7.75-15	94P / TL	BFGoodrich	5.50-18	80P; 84P / TT	Goodrich; Firestone
5.20-13	70P / TL	Dunlop; Firest.	7.75-]5	94P / TL	Firestone	5.50-18	NA / TT	Michelin
5.60-13	75S / TL	BFGoodrich	8.00-15	92H / TL	Dunlop	5.50/6.00-18	76H/TT	Dunlop
5.60-13	75P / TL	Firestone	8.15-15	97P / TL	BFGoodrich	6.00/6.50-18	87P; 98P; 90P/TT	Firest.; Dunlop; Excels.
5.90-13	79S / TL	BFGoodrich	8.15-15	98H/TT	Dunlop	6.00/6.50-18	NA / TT	Michelin
6.00-13	80P / TL	BFGoodrich	8.15-15 6PR	98H / TL	Avon	7.00-18	102P; 107P/TT	Dunlop; Firestone
6.40-13	81S/TL	BFGoodrich	8.20-15	99H/TT	Dunlop	3.50-19	56L; 53P/TT	Avon; Dunlop
6.40-13	82S / TL	Dunlop	8.20-15	103P/TL	BFGoodrich	3.50/4.00-19	56P/TT	Excelsior
7.25-13	90S / TL	BFGoodrich	8.20-15	103P/TL	Firestone	4.00-19	64S; 60P / TT	Avon; Dunlop
7.50-13	85H / TL	Dunlop	8.20-15 6PR	102H/TL	Avon	4.00/4.50-19	NA / TT	Michelin
5.00/5.20-14	74P/TT	Excelsior	8.55-15	99P / TL	BFGoodrich	4.50-19	69P; 71 N / TT	Dunlop; Excelsior
5.20-14	74P / TL	Dunlop	8.90-15	n OS / TT	Dunlop	4.50-19	69P; 71N/TT	Firestone; Waymaster
5.60-14	76P / TL	BFGoodrich	8.90-15	108P/TL	Firestone	4.75/5.00-19	76N; 76P / TT	Waymaster; Dunlop
6.95-14	76P / TL	BFGoodrich	9.00-15	103P/TL	BFGoodrich	4.75/5.00-19	76N; 76P / TT	Goodr; Excels.; Firest.
6.95-14	86P / TL	BFGoodrich	4.50/4,75-16	66P/TT	Firestone	4.75/5.00-19	NA / TT	Michelin
7.00-14	90P / TL	BFGoodrich	5.00/5.25-16	74P; 76L; 83P / TT	Dunlop; Excelsior	5.00-19	69H/TT	Dunlop
7.35-14	90P / TL	BFGoodrich	5.00/5.25-16	74P; 76L; 83P / TT	BFGoodrich	5.25/5.50-19	82N; 90P / TT	Waymaster; Excelsior
7.50-14	96P / TL	BFGoodrich	5.00/5.25-16	74P; 76L; 83P / TT	Avon; Firestone	5.25/6.00-19	o.A./TT	Michelin
7.50-14	95P / TL	Firestone	5.50-16 (TT)	74P; 76P; 80L; 82P	BFGoodrich	5 25/5.50/6.00-19	84P / TT	Dunlop
7.75-14	96P / TL	BFGoodrich	5.50-16 (TT)	74P; 76P; 80L; 82P	Firestone; Dunlop	5.50-19	82P/TT	BFGoodrich
7.75-14	95P / TL	Firestone	5.50-16 (TT)	74P; 76P; 80L; 82P	Avon; Excelsior	5.50/6.00-19	82P/TT	Firestone
8.00-14	97P / TL	BFGoodrich	6.00-16 (TT)	80H; 86L; 92P; 95P	Dunlop; Avon	6.00-19	77H/TT	Dunlop
8.25-14	97P/TL	BFGoodrich	6.00-16 (TT)	80H; 86L; 92P; 95P	Firest.; Excelsior	6.00/6.50-19	93P/TT	BFGoodrich
8.50-14	99P / TL	BFGoodrich	6.00-16 6PR	89H/TT	Avon	6.50-19	84H; 93P / TT	Dunlop; Firestone
8.55-14	99P / TL	BFGoodrich	6.00-16	92P/TT	BFGoodrich	6.50/7.00-19	102P/TT	Dunlop
8.85-14	101P/TL	BFGoodrich	6.50-16	87P; 96P; 102P/TT	BFGoodrich	7.00-19	88H; 96P / TT	Dunlop; Firestone
9.00-14	101 P /TL	BFGoodrich	6.50-16	87P; 96P; 102P/TT	Excelsior; Firest.	4.50/4.75/5.00-20	77P / TT	Dunlop
9.50-14	104P/TL	BFGoodrich	6.70-16	85P; 89H / TT	Dunlop; Firestone	4.75/5.00-20	77P; 86P / TT	Goodr.; Excels.; Firest.
4.00/4.25-15	76P/TT	Excelsior	7.00-16	91H; 102P/TT	BFGoodrich	5 25/5.50/6.00-20	93P / TT	Dunlop
4.25/4.40-15	o.A./TT	Michelin	7.00-16	91H;102P/TT	Dunlop; Firestone	5.50-20	83P/TT	BFGoodrich
4.00/4.25-15	74N/TT	Waymaster	7.50-16	98P / TL	Dunlop	6.00-20	91P; 92P/TT	Goodrich; Firestone
4.50-15	74N/TT	Waymaster	7.50-16	98P; 101P/TT	Firest.; Goodrich	6.50-20	106P/TT	Firestone
5.00-15	68P / TL	BFGoodrich	8.25-16	105P/TT	BFGoodrich	6,50/7.00-20	102P / TT	Dunlop
5.50-15	72P/TT	Excelsior	4.00-17	69P/TT	Excelsior	6.50/7.00-20	a A. / T T	Michelin
5.60-15	78P / TL	Firestone	4.00/4.25-17	77N/TT	Waymaster	7.00-20	99P/TT	BFGoodrich
5.90-15	82P / TL	BFGoodrich	4.50-17 (TT)	65P; 69P; 76P; 77L	Firest.; Excelsior	4.40/4.50-21	75P/TT	Goodrich; Firestone
5.90-15	79H/TL	Dunlop	4.50-17 (TT)	77L	Dunlop; Avon	4.50/4.75-21	78P; 75P / TT	Dunlop; Excelsior
6.00-15	85P / TL	BFGoodrich	4.75/5.00-17	78P / TL	Dunlop	5.00/5.25-21	84P/TT	Dunlop; Excelsior
6.40-15	86S; 86P / TL	Avon; Goodrich	4.75/5.00-17	77N; 79P / TT	Waymaster; Excels.	5.25-21	84P/TT	Firestone
6.40-15	87P / TL	Firestone	5.25/5.50-17	78P; 88N / TT	Firest.; Waymaster	5.50/6.00-21	NA / TT	Michelin
6.40-15	86H; 78P / TT	Dunlop; Excels.	5.25/5.50-17	88P; 80P / TT	Dunlop; Excelsior	6.00-21	91P; 93P/TT	Goodrich; Dunlop
6.00/6.40-15	86P / TL	Dunlop	5.50-17	72H; 78P / TT	Dunlop; Goodrich	7.00-21	103P/TT	Dunlop
6.70-15	92S; 92P / TL	BFGoodrich	6.00/6.50-17	91P; 98P/TT	Firestone; Dunlop	7.00-21	o. A. / TT	Michelin
6.70-15	88H; 93P / TL	Dunlop; Firest	6.50/7.00-17	NA / TT	Michelin	6.00-22	91P/TT	BFGoodrich
6.70-15	82P/TT	Excelsior	7.00-17	93P; 102P/TT	Firestone ; Dunlop	5.00-23	81P/TT	Excelsior
6.70-15 6PR	91H/TL	Avon	7.50-17	98P/TT	BFGoodrich	5.00-24	81P; 82P/TT	Goodrich; Excelsior
7.00-15	98P / TL	BFGoodrich	4.50-18	71 P; 69P / TT	Dunlop; Excelsior	38x7 (7.50-24)	113P/TT	BFGoodrich
7.00-15	95P / TL	Firestone	4.75/5.00-18	80N; 80P /TT	Waymaster; Dunlop	40x8 (9.00-24)	125P / TT	BFGoodrich

TT stands for tube-type (with inner tube), TL stands for tubeless; NA = data not available

The right choice for a Jaguar XK: the Michelin Pilote X

Michelin XAS: popular in the Sixties

A classic from Holland: the Vredestein Sprint

Avon: OEM supplier to Rolls-Royce too

Sporting: the XWX from Michelin, the market leader

Michelin TRX: new in the Seventies

RADIAL TYRES

Size	Load rating/ inner tube	Brands available
125R12	62S / TL	Pirelli
165/70R10	68S / TL	Dunlop
135/80R13	70 T / TL	Kumho
145/80R13	75 T / TL	Kumho
145SR10	68S / TL	Dunlop
155/80R13	79T/TL	Kumho
155/70R13	75 T / TL	Kumho
165/80R13	83T / TL	Kumho
165/70R13	83T / TL	Kumho
155R13	78S / TL	Dunlop
155HR13	78H/TT	Michelin
165R13	82S / TL	Dunlop
165HR13	82H/TT	Michelin
175/70R13	82S/TL	Dunlop; Kumho
175/65R13	80T / TL	Kumho
185/70R13	85S / TL	Dunlop
185HR13	88H/TT	Michelin
185/70VR13	86V / TL	Michelin
205/70VR13	91V/TL	Michelin
6.40/7.00SR13	87S/TT	Michelin
7.25R13	90S/TT	Michelin
165/70R14	81T/TL	Kumho
165HR14	84H/TT	Michelin
175R14	88H / TL	Vredestein
175/70R14	84H / TL	Kumho
185R14	90H; 91H/TL	Dunl.; Vredest.
185HR14	90H / TL	Mich.; Vredest.
185/70R14	88T; 88H / TL	Kumho; Mich.
195/70R14	91H/TL	Kumho
205VR14	89W / TL	Michelin
205/70R14	95V / TL	Vredestein
205/70VR14	89W / TL	Michelin
215/70VR14	92W / TL	Michelin
125R15	68S / TL	Michelin
135SR15	72S / TL	Michelin
135R15	72Q / TL	Michelin

Size	Load rating/ inner tube	Brands available
145SR15	78Q / TL	Michelin
155R15	82S / TL	Vredestein
155HR15	82H/TT	Michelin
155SR15(M+S)	82Q / TL	Vredestein
165SR15	86S / TL	Michelin
165R15	86H / TL	Avon; Vredestein
165SR15(M+S)	86Q / TL	Vredestein
165HR15	86H/TT	Michelin; Vredestein
180HR15	89H/TT	Michelin
185R15	91H; 92V/TL	Dunlop; Avon
185R15	93H / TL	Vredestein
185HR15	93H / TL	Michelin
185VR15	93V / TL	Michelin
185VR15	91V/TT	Dunlop
185/70R15	91H/TL	Vredestein
185/70VR15	89V / TL	Michelin
ER70R15	93H / TL	Dunlop
205R15	97H / TL	Avon
205/70R15	95V; 96V / TL	Avon; Vredestein
205/70VR15	90W; 95W / TL	Michelin; Pirelli
215/70VR15	90W / TL	Michelin
225/70R15	100V/TL	Avon
225/70VR15	92W / TL	Michelin
235/70R15	101V/TL	Avon
235/70HR15	101H/TL	Michelin
255/45VR15	93W / TL	Michelin
175R16	98N/TT	Dunlop
185R16	92S/TT	Michelin
185HR16	93H/TT	Vredestein
225/50ZR16	92Y/TT	Michelin
255/50ZR16	100Y / TL	Michelin
5.50R16(165R16)	88H / TL	Avon
5.50R16	84H/TT	Michelin
6.00R16	88W/TT	Michelin
6.70R16	97V/TL	Avon
—	—	—

RADIAL TYRES (metric)

Size	Load rating/ inner tube	Brands available
190/55VR340	881V/TL	Michelin
170/65R365	82H / TL	Michelin
220/55VR365	92V / TL	Michelin
190/65HR390	89H / TL	Michelin
210/55VR390	91V/TL	Michelin
220/55VR390	93W/TL	Michelin
200/60VR390	90V / TL	Michelin
240/55VR390	89W / TL	Michelin
125R400	69S/TT	Michelin
135R400	73S/TT	Michelin
145R400	79S/TT	Michelin
155R400	83S/TT	Micheltn
165R400	87S/TT	Michelin
185R400	91S/TT	Michelin
17R400	103M/TT	Michelin
19R400	112; HON/TT	Firestone
240/45VR415	94W / TL	Michelin
240/55VR415	94W / TL	Michelin
280/45VR415	91Y/TL	Michelin

TT stands for tube-type (with inner tube), TL stands for tubeless; NA = data not available

Above: Since 2007, the Dunlop Sport D8 for the 300 SL 'Gullwing' 300 SL has again become available.

Far left: Curing a tyre: over time, tyre production has become increasingly mechanised.

Left: in the early days, pulling the rubber onto the rims was hard work.

Opposite page: The relationship between lovely legs and car tyres was already established in advertising in the 1920s.

recently experienced considerable growth. That should particularly please drivers of Cobras, Lamborghinis, Maseratis and de Tomasos. The British supplier, The Blockley Tyre Company mentioned earlier, also specialises in tyres for historic motorsport. If you still cannot find the right tyre size for your car, get in touch with some of the companies mentioned earlier in the chapter, or find out more from the individual manufacturers.

Whitewall radial tyres (tubeless)

Size	Load rating	Brands available
P155/80R13	79S	BFGoodrich, Maxxis
P165/80R13	83S	BFGoodrich, Maxxis
P175/80R13	86S	BFGoodrich, Maxxis
P185/80R13	89S	BFGoodrich, Maxxis
185R14	90H	Phoenix
P205/75R14	95S	BFGoodrich
P185/70RU	87S	BFGoodrich
P215/70R14	96S	BFGoodrich
P225/70R14	98S	BFGoodrich
165R15	86S	BFGoodrich, Coker
P205/75R15	97S	BFGoodrich
P215/75R15	99S	BFGoodrich
P225/75R15	102S	BFGoodrich
P235/75R15	105S	BFGoodrich
P185/70R15	88S	BFGoodrich
P215/70R15	97S	BFGoodrich
P225/70R15	100S	BFGoodrich
P235/70R15	102S	BFGoodrich
P255/70R15	107S	BFGoodrich
P285/70R15	115S	BFGoodrich

Radial tyres for American cars (tubeless)

Size	Load rating	Brands available
P175/70R13	782S	BFGoodrich
P185/70R14	87S	BFGoodrich
P185/75R14	89S	Maxxis
P195/60R14	85S	BFGoodrich
P195/70R14	90S	BFGoodrich
P195/75R14	92S	Maxxis
P205/70R14	93S	BFGoodrich; Maxxis
P205/75R14	95S	Maxxis
P215/60R14	91S	BFGoodrich
P215/70R14	96S	B- F. Goodrich; Maxxis
P225/60R14	94S	BFGoodrich
P225/70R14	98S	BFGoodrich
P235/60R14	96S	BFGoodrich
P245/50R14	93S	BFGoodrich
P245/60R14	98S	BFGoodrich
P265/50R14	98S	BFGoodrich
P155/80R15	83S	BFGoodrich
P195/60R15	87S	BFGoodrich
P205/55R15	87S	BFGoodrich
P205/60R15	89S	BFGoodrich
P205/70R15	95S	Maxxis
P205/75R15	97S	Maxxis
P215/60R15	92S	BFGoodrich
P215/65R15	95S	BFGoodrich
P215/70R15	97S	BFGoodrich; Maxxis

Size	Load rating	Brands available
P215/75R15	100S	Maxxis
P225/60R15	95S	BFGoodrich
P225/70R15	100S	BFGoodrich; Maxxis
P225/75R15	102S	Maxxis
P235/60R15	98S	BFGoodrich
P235/70R15	102S	BFGoodrich
P235/75R15	105S	Maxxis
P245/60R15	100S	BFGoodrich
P255/60R15	102S	BFGoodrich
P255/70R15	108S	BFGoodrich
P275/50R15	101S	BFGoodrich
P275/60R15	107S	BFGoodrich
P285/70R15	115S	BFGoodrich
P295/50R15	105S	BFGoodrich
P195/50R16	83S	BFGoodrich
P205/50R16	86S	BFGoodrich
P205/55R16	89S	BFGoodrich
P225/50R16	91S	BFGoodrich
P245/50R16	96S	BFGoodrich
P265/50R16	101S	BFGoodrich
P265/60R16	106S	BFGoodrich
P275/65R16	ms	BFGoodrich
P285/60R16	1115	BFGoodrich
P295/50R16	107S	BFGoodrich

Maxxis has replaced Cooper as the supplier of tyres with complete whitewalls or bands.

Act X: at the end of the 1940s at the end of the 1940s, Michelin's radial tyre made its triumphant appearance.

Citroën was the first car-maker to fit the Michelin X; other manufacturers soon followed suit.

Excelsior competition tyres are of interest for cars capable of more than 124mph (200km/h).

The tread patterns follow the original moulds.

The Avon CR 6 ZZ provides decent rubber for high-performance sports cars.

Excelsior competition tyres approved for road use (tubeless)

Size	Speed rating	Size	Speed rating
5.50-18	H	5.50-17	V
4.00-19	H	6.00/6.50-17	V
4.50-19	H	7.00/7.50-17	V
5.00-19	H	4.75/5.00-18	V
5.50/6.00-19	H	5.50-18	V
6.50/7.00-20	H	6.00/6.50-18	V
5.50/6.00-21	H	7.00-18	V
7.00-21	H	4.00-19	V
5.00-15	V	4.50-19	V
5.50-15	V	5.00-19	V
6.00-15	V	5.50-19	V
6.50-15	V	6.00-19	V
7.00-15	V	6.50-19	V
5.50-16	V	7.00-19	V
6.00-16	V	6.00-20	V
6.50-16	V	6.50/7.00-20	V
7.00-16	V	5.50/6.00-21	V
7.50-16	V		

Avon competition tyres

Size	Load rating
165/70R10	72H
175/70R13	82H
185/70R13	86H
185/70R14	88H
175/70R15	86H
185/70R15	89H
205/70R15	96V
215/60R15	99V
215/70R15	98V
225/65R15	99V
245/60R15	101V
275/55R15	104V
295/50R15	108V

Racing successes always made for compelling advertisements: the world champion Alberto Ascari advertises Pirelli's tyres.

LOAD RATINGS

Rating	kg per tyre	Rating	kg per tyre	Rating	kg per tyre	Rating	kg per tyre	Rating	kg per tyre	Rating	kg per tyre
50	190	65	290	80	450	95	690	110	1060	125	1650
51	195	66	300	81	462	96	710	111	1090	126	1700
52	200	67	307	82	475	97	730	112	1120	127	1750
53	206	68	315	83	487	98	750	113	1150	128	1800
54	212	69	325	84	500	99	775	114	1180	129	1850
55	218	70	335	85	515	100	800	115	1215	130	1900
56	224	71	345	86	530	101	825	116	1250	131	1950
57	230	72	355	87	545	102	850	117	1285	132	2000
58	236	73	365	88	560	103	875	118	1320	133	2060
59	243	74	375	89	580	104	900	119	1360	134	2120
60	250	75	387	90	600	105	925	120	1400	135	2180
61	257	76	400	91	615	106	950	121	1450	136	2240
62	265	77	412	92	630	107	975	122	1500	137	2300
63	272	78	425	93	650	108	1000	123	1550	138	2360
64	280	79	437	94	670	109	1030	124	1600	139	2430

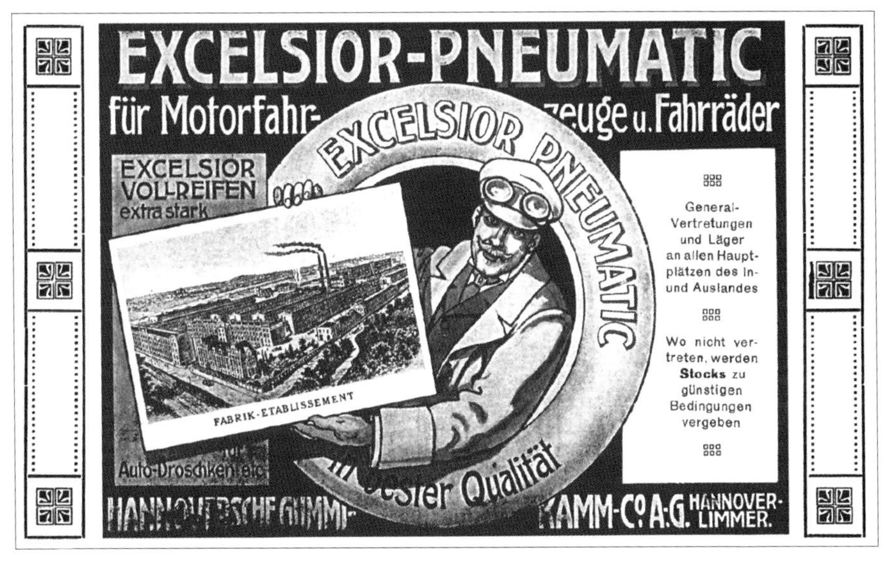

Excelsior was originally the brand name used by the Hanover Rubber Works – today it is applied to competition tyres from MOR.

SPEED RATING

Rating	approved for use up to
A1-A8	from 5 to 40kph (3-25mph) in 5kph increments
B	50km/h = 31mph
C	60km/h = 37mph
D	65km/h = 40mph
E	70km/h = 43mph
F	80km/h = 50mph
G	90km/h = 56mph
J	100km/h = 62mph
K	110km/h =68mph
L	120km/h =75mph
M	130km/h = 81 mph
N	140km/h = 87mph
P	150km/h = 93mph
Q	160km/h = 99 mph
R	170km/h = 106mph
S	180km/h = 112mph
T	190km/h = 118mph
U	200km/h = 124mph
H	210km/h = 130mph
V	240km/h = 149mph
VR	over 130mph
W	270km/h = 168mph
ZR	over 149mph
Y	300km/h = 186mph

Variations on a theme: classic tyres are available in most common sizes.

Conversion table when changing different series of tyres*

83-Series before 1964	83-Series from 1965-72	82-Series from 1970	Alphanumeric 78-Series	75-Series from 1982	70-Series from 1983
5.90-13	6.00-13	165-13	A78-13	165/75R 13	175/70 R13
6.40-13	6.50-13	185-13	B78-13	185/75R 13	195/70R13
5.90-14	6.45-14	165-14	B78-14	165/75 R14	175/70R14
6.50-14	6.95-14	185-14	C78-14	185/75R14	195/70R14
7.50-14	7.75-14	195-14	F78-14	195/75R 14	205/70R 14
8.00-14	8.25-14	205-14	G78-14	205/75 R14	215/70R 14
8.50-14	8.55-14	215-14	H78-14	215/75R 14	225/70R 14
5.90-15	6.00-15	165-15	A78-15	165/75R 15	175/70R 15
6.40-15	7.35-15	185-15	E78-15	185/75 R 15	195/70R 15
6.70-15	7.75-15	195-15	F68-15	205/75R15	215/70R 15
7.10-15	8.25-15	205-15	G78-15	215/75R 15	225/70 R 15
7.60-15	8.55-15	215-15	H78-15	225/75R 15	235/70R 15
8.20-15	9.00-15	235-15	L78-15	235/75R 15	255/70 R15

* = for guidance only, subject to check of steering clearance and suspension travel (source: Möller Oldtimerreifen)

Wheels
Restoring wire wheels

Wire wheels are surely among the most beautiful details on a classic car. But lovely as the fine wire spokes are, they can cause plenty of problems when you come to restore them. We will show you, step-by-step, what you can do yourself, and which jobs are best left to professionals.

Many wire wheels sold today come from India, which, on the one hand explains their relatively inexpensive prices, and, on the other, the substandard quality of the materials and finish in many cases.

Wire-wheel specialists generally recommend that with new wheels all the spokes should be re-tensioned after a few hundred miles, especially on wheels made in India. After 100-200 miles, all the spokes and nipples will have settled in position; if you keep on driving without thinking or paying attention to the wheels, you risk breaking the spokes or causing an irreparable imbalance in the wheel. The rest is a matter of looking after them, and we will come to that later.

In terms of quality, you will do much better to choose established brand-name products. But even so, 300 miles (500km) after a complete restoration, all the spokes should be thoroughly inspected and re-tightened; after that, you should enjoy peace of mind for a very long time. The chrome and polished surfaces on the wheel rim and hub especially are far superior to the Indian products. The nipples and spokes are also made from top-quality materials. Today, aluminium is used almost exclusively for wheel rims, whether for cars or motorcycles. Borrani now supplies only car wheels; its once dominant position in the motorcycle market having been claimed by the Spanish manufacturer Akront. British wheels are available under the Alu-Rim brand name. For older vehicles, nearly all sizes (for cars and motorbikes) are available in steel. Nowadays, many sizes are also sold in stainless steel. For a long time, however, this material was considered technically rather unreliable. The so-called 'thick-ended' spokes in particular were unavailable in stainless steel for a long while.

We interviewed a German manufacturer who offers superb-quality stainless-steel wheels in nearly all sizes. "I don't hesitate to recommend to nearly every classic car owner that they should have their wheels restored with polished stainless-steel spokes. Now that there are datasheets and strength test certificates for the materials used and the production process, the TÜV testing authorities have nothing against them either," the specialist observes. As well as precise data for the alloy used, these documents provide measurements for their tensile strength, breaking strength and fatigue strength. All the stainless-steel spokes used by most firms comply with the DIN 74371 (Part 1) or ETRTO (European Tyre and Rim Technical Organization) standards, or the requirements of the US Department of Transportation. "Unfortunately, there are a huge number of poor-quality products of dubious origin on the market, which use metal better suited to kitchen sinks or watering cans. The problem is that it is practically impossible for the layman to tell these spokes apart from good-quality products.

Plastic-coated hubs and rims, stainless-steel spokes and chrome-plated brass nipples – the ideal combination to last for eternity, which can be used on both cars and motorcycles.

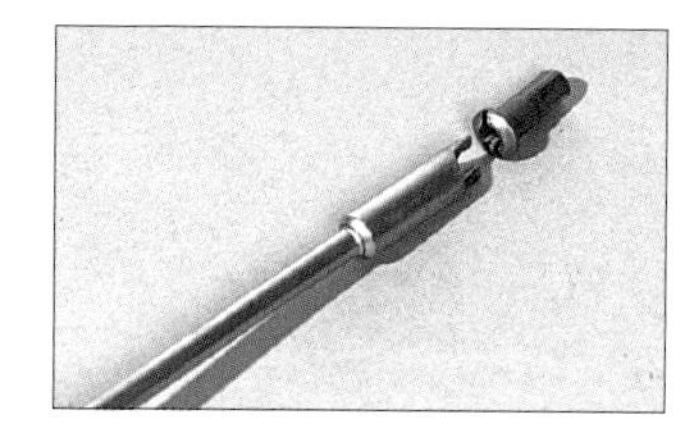

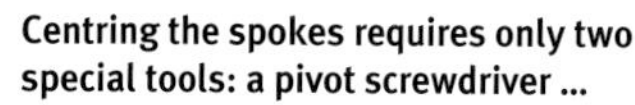

Centring the spokes requires only two special tools: a pivot screwdriver ...

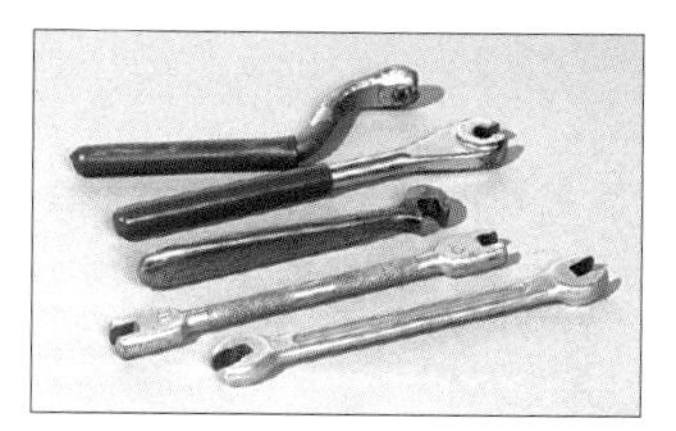

... and a spoke spanner. Professionals favour the three-sided version (with the blue grip).

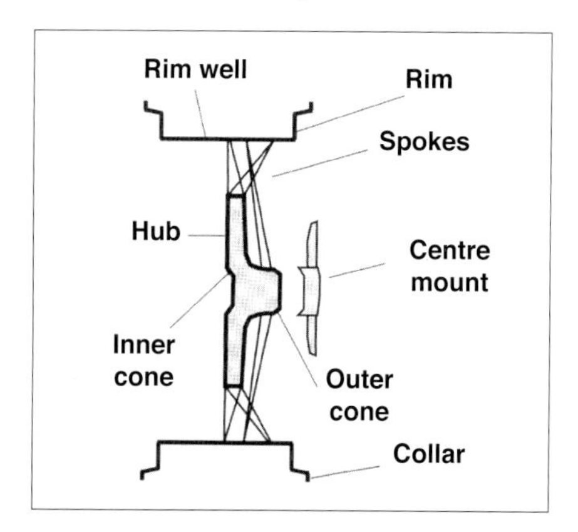

A real classic: the drawing shows how a Borrani wheel is typically assembled.

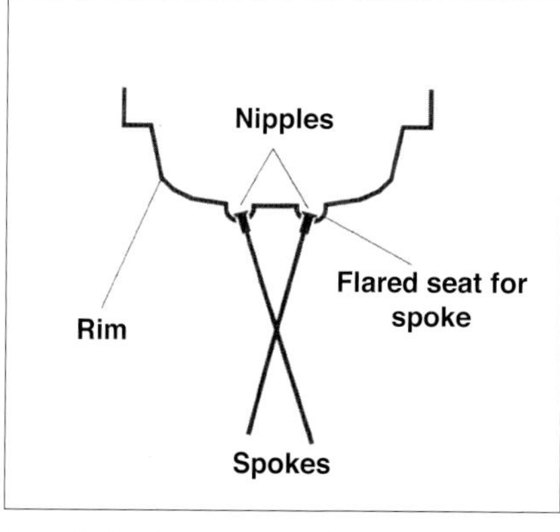

These 'bulges' serve as seats for the spokes.

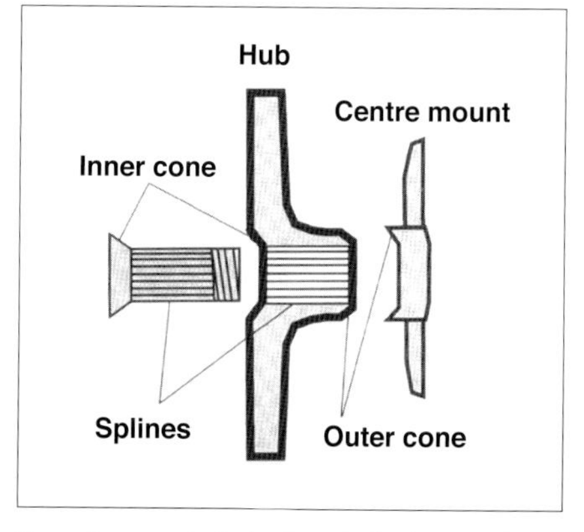

For centre-mounted wheels, only the inner and outer cones are used to centre them.

Originally, rulers were used. Here the exact position of the hub is measured.

The technician measures the diameter of the sample spoke, which has been carefully removed.

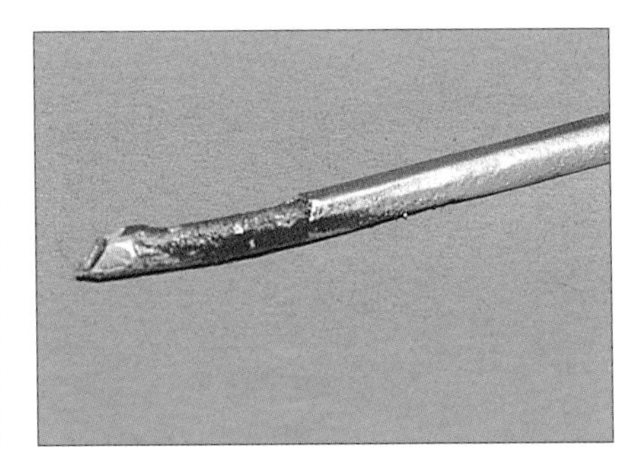

Off with the paint! Otherwise, the old layers of paint will affect the measurement readings.

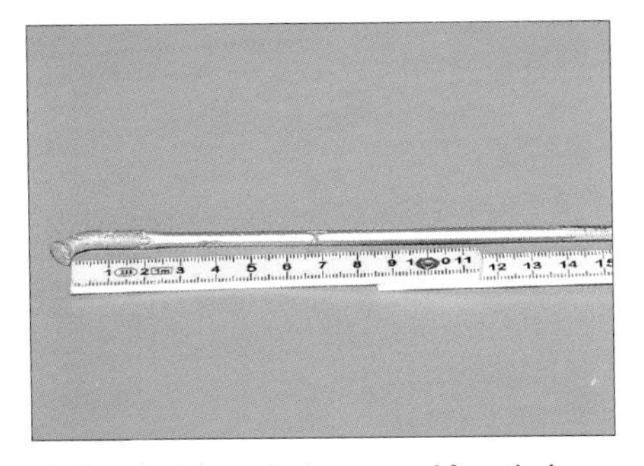

The length of the spoke is measured from the inner edge of the spoke head to the end of the thread.

A glance at the outside of the rim will reveal any spokes that are standing proud of the nipples.

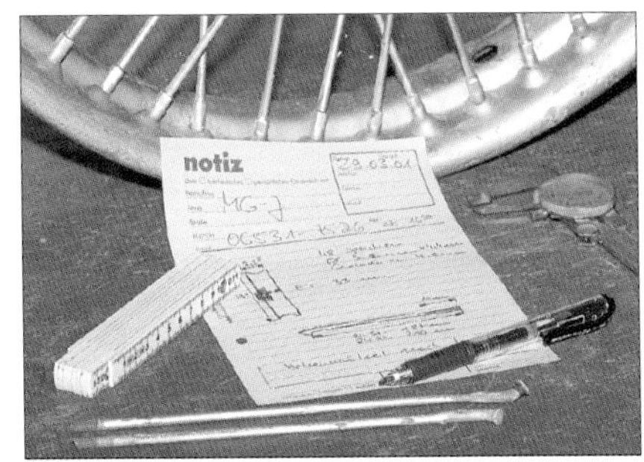

All the measurements are noted and thus provide a complete record for the wheel.

"To play it safe, don't hesitate to ask to see the technical documentation from a wheel builder you trust," the expert advises.

Another technology to have made inroads into the classic car market in recent years is that of plastic coating. Specialist firms can now cover pretty much any component with a very strong plastic coating. First of all, a layer of the appropriately coloured powder is applied electrostatically to the part, then baked on at very high temperature. Teething troubles, such as the coating flaking off, or a limited selection of colours are now a thing of the past. Since the process today makes it possible to apply a good base layer, the corrosion protection is also considered to be reliable.

We got going with our test wheels, from a 1933 MG J2: with their size of 19 x 2.5 inches, they could just as well have come off a motorcycle. The MG's wheels, however, had one special feature: the spokes were offset, that is, the nipples were not located in the middle of the rim band, but on the edge. This allowed the English manufacturer to save a couple of centimetres in their construction and to have a positive effect on the steering geometry.

You can start to save money when you deliver the wheels. If you don't take the wheels apart yourself, you'll end up paying a bit more. Any further work you may undertake yourself should be precisely agreed with the specialist firm you use. If you have the confidence to do the job yourself, you can, of course, just entrust the spokes and nipples to them. When it comes to centring the wheels, however, you should let the professionals take over.

Before dismantling the wheels, take a note of how they are constructed. Professionals begin with the wheel offset. This term refers to the location of the hub in relation

Bolt cutters to the ready! Somehow we imagined that taking our good old 'wires' apart would have been a more delicate operation.

Done already: the spokes and nipples are only fit for scrap now. The puzzle for advanced players starts here ...

Before the shot-blasting and paint removal, you must get rid of any remaining grease altogether.

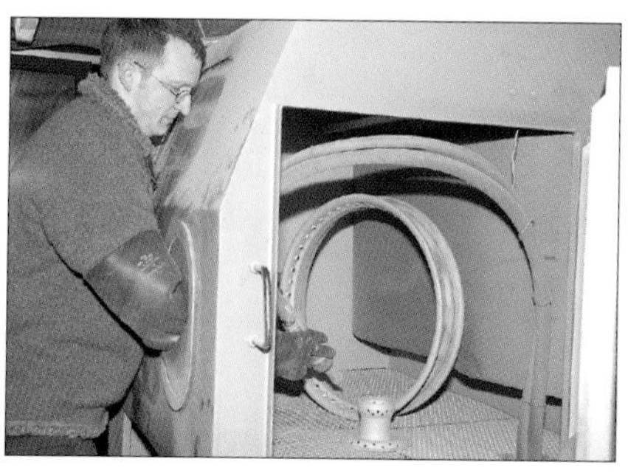

As preparation for plastic coatings, it is much better to treat the surface with a coarse abrasive.

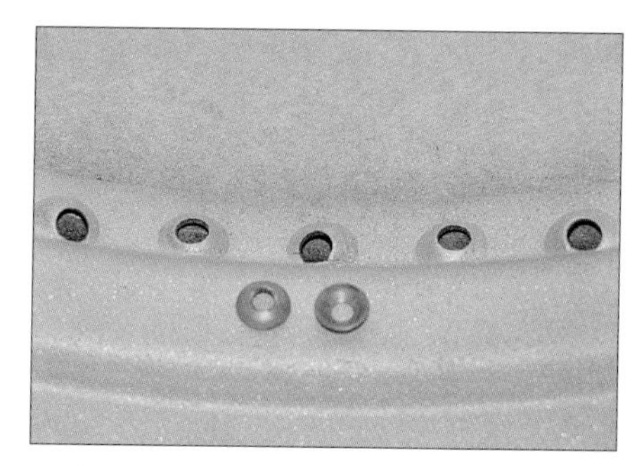

Cracks around the spoke holes? These little domed washers are often a knight in shining armour.

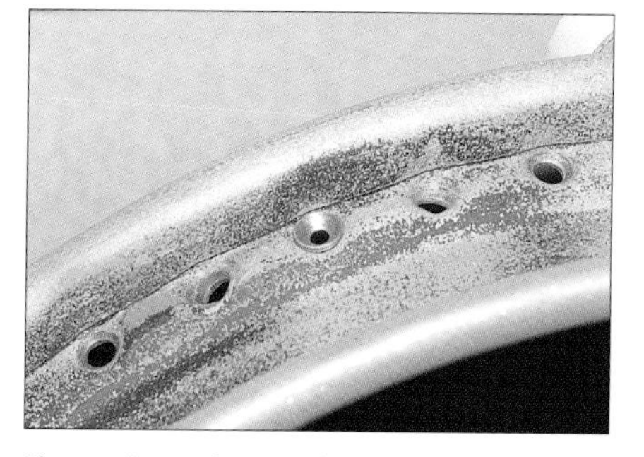

They are inserted on top of the hole and then brazed in place.

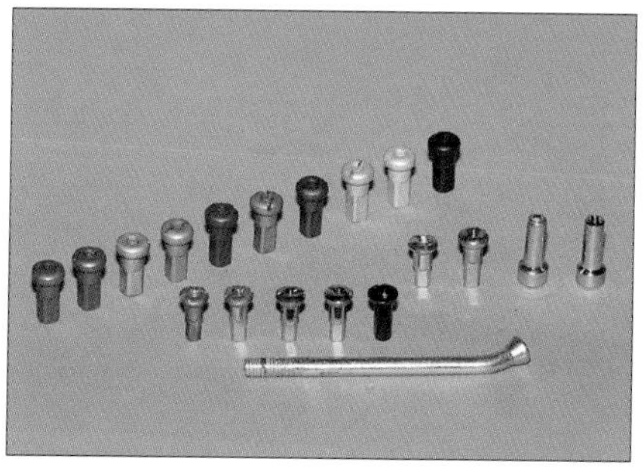

What colour would you like? Nipples are available to suit (almost) every taste.

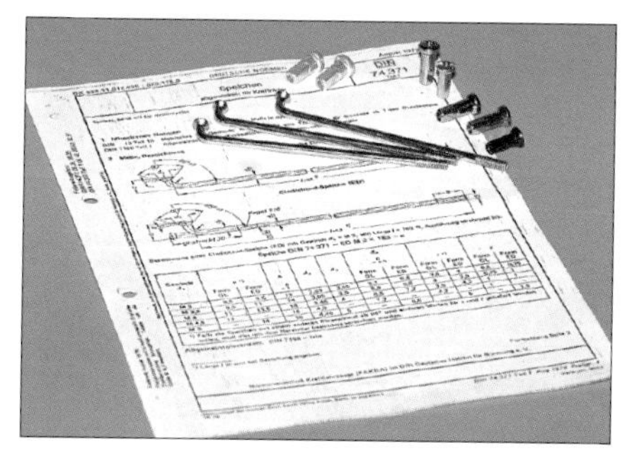

Vehicle examiners love a comprehensive datasheet.

An unrestricted choice: firms with a good stock carry nearly every type of stainless-steel spoke.

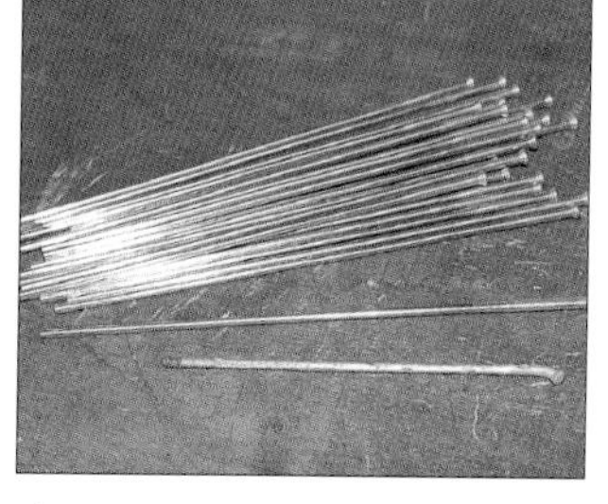

The new spoke rods for our wheel: the original spokes in the foreground served as a template.

to the edge of the wheelrim. Our wheels had a total of 48 spokes: 24 short and 24 long. The expert removes one of each as a sample, in order to measure their length and diameter. A few layers of paint, which have built up over the years, must first be removed. The outside edge of the wheel warrants a closer look. If the spokes are sticking out beyond the nipple, they have increased in length over time as the result of frequent re-tensioning. This increase must be deducted when the new spokes are cut to length, otherwise they could damage the rim or the inner tubes when the tyres are fitted.

Only when all the data has been collected and noted can the wheel be fully disassembled. Since the old spokes are no longer of any use, they can be broken off with bolt cutters to save time. Any residual grease or copper paste should now be cleaned off the wheel hub, before the parts are taken to a specialist to have the paint stripped off in a dip tank. If several coats of paint have to come off, the time needed to blast off the paint would otherwise be too great. What's more, shot-blasting can leave the surface too rough. If the shot becomes contaminated with paint, it will have to be replaced and disposed of at additional expense.

Depending on the desired surface finish, however, the parts still have to undergo a further fine blasting treatment. If they are going to be painted, glass bead blasting is the correct method to use. If you plan to coat them in plastic, the outer surface can – and indeed should – be somewhat rougher: corundum or similar coarse abrasives are the right products to use in this case. To ensure that they are protected against corrosion, these wheel components are additionally treated with phosphate. For wheels with a central mounting, however, neither the cone nor the splines should be shot-blasted. This section of the wheel should be carefully covered and subsequently kept free from paint or the plastic coating; if not, a good fit cannot be guaranteed.

Unfortunately, it is often only after the wheels have been shot-blasted that damage to the flared areas around the spoke holes can become apparent. These are the slight bulges where the nipple heads are located. "In principle, these are potential breaking points, which make it impossible to go on using the wheel with a clear conscience," the expert warned us. A brand-new rim for our MG is available at a cost of at least •x220. To this should be added nearly •x135 per wheel to punch the spoke holes. A tidy sum for five wheels ...

This problem led to the development of domed washers or ferrules, which can be inserted into the damaged section around the spoke holes from the exterior and brazed in place. You can then go on using the wheel. This time-consuming manual work is only worthwhile, however, if there are only a few cracks. If the rim is damaged all the way round, a new wheel will

This so-called 'elbow' is produced with the help of a special bending machine.

Naturally, the radius and also the angle of the elbow must match the original.

Hand work is called for: each rod must be cut to length individually.

Milling rather than cutting – a job for professionals. Different thread sizes can be produced.

Once the desired length has been reached, the machine automatically removes the thread.

A test nipple shows whether the thread is satisfactory and turns smoothly.

be more cost-effective. Unfortunately, this approach has only worked for wheels made from steel. The next step is to select suitable nipples and rods for the spokes.

The ideal combination has proven to be stainless-steel spokes and nickel-plated brass nipples. Steel nipples are also available, but these have a slight tendency to chafe against the threads. "After many years, the brass nipples still turn on the spokes as smoothly as a knife going through butter," the specialist lets on. Wheel builders today offer their customers a wide range of different nipples, made from aluminium (anodised in multiple colours), brass, steel or stainless steel. Even customers who would like gold-plated nipples can be catered for. Next, the rods are given their so-called 'elbow' on a machine powered by compressed air. These 'elbows' are bent in place just before the spoke head, exactly as with the original sample. An incorrect elbow angle and the wrong bending radius can lead to the metal breaking. Exact machine settings and subsequent visual checks are essential. Finally, the rods are cut to the prescribed length using a cutting machine with a length stop.

All that is missing now is the thread, which is milled on a special machine. Time and again, amateur mechanics ask why a spoke they have already replaced has broken yet again? In cases such as these, the threads have generally been cut, weakening the metal and thereby creating a dangerous potential breaking point. For this reason alone, amateurs are rarely successful using blank rods to make their own spokes.

When boring holes in a new alloy wheel, it is essential to pay attention to the exact angle!

The importance of ensuring that the angle and seating for the spoke holes are correct soon becomes clear when ...

... you compare the completely different exit angles of these two wheels.

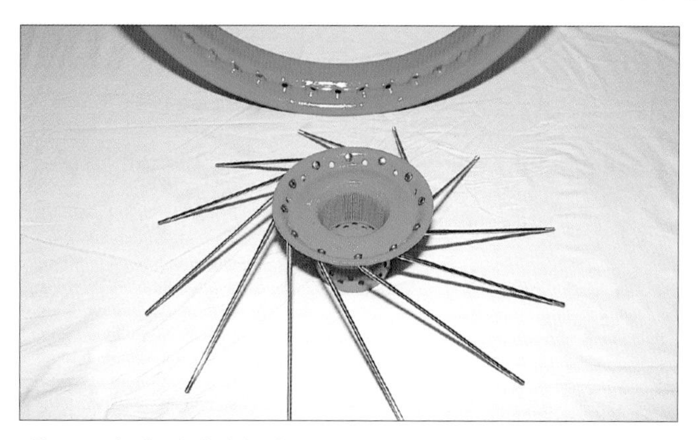

Off we go: the first half of the short spokes are inserted.

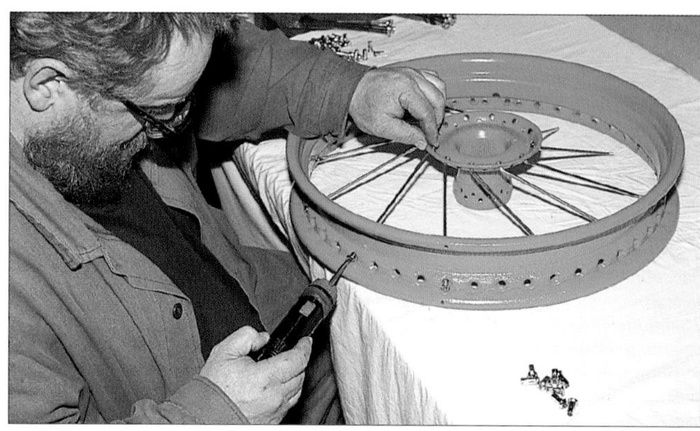

The professional wheel builder tightens the nipple quite loosely to begin with.

The thirteenth short spoke, and all those that follow it, are laid crosswise.

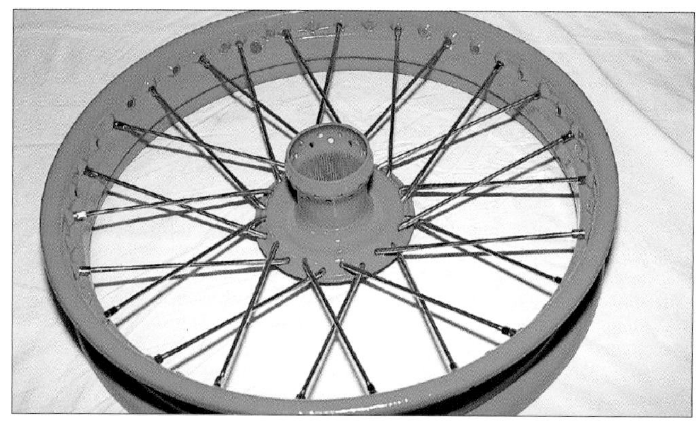

The 24 short spokes have now been assembled, giving a uniform appearance.

The long spokes are attached to the outer hub ring.

The specialist now tightens the nipples so that the threads are hidden.

After our wheel rims and hubs came back from being plastic-coated, we assembled all the parts, and could begin lacing the spokes. With a relatively simple pattern such as that on the MG, the expert considered this work could be undertaken by an experienced amateur mechanic.

The 24 short spokes run from the inner hub ring to the rim, while the 24 longer spokes are attached to the outer hub ring. Of these 24 spokes, 12 always run in one direction and 12 in the opposite direction, so that they cross over each other. If you have good documentation and photographs to refer to, you should have no problems. "In the case of old Borrani wheels, which can have up to six different spoke lengths and a very complicated lacing pattern, however, I would strongly discourage amateur mechanics from trying to assemble them," the specialist warned.

In order to ensure that the wheel would run reasonably true, even at this stage, the technician unscrewed all the nipples to the point that they just covered the spoke threads. At this point, they were all relatively slack. Next, each nipple was given a further half turn, whole turn or two turns, depending on how much slack could still be felt in the complete assembly. Gradually, all the spokes responded to gentle pressure and the hub was stabilised. The wheel was then transferred to the test bench, to determine the lateral and radial run-out.

Depending on the values shown on the precision measuring equipment, the expert released the tension on some spokes and increased it on others, which he

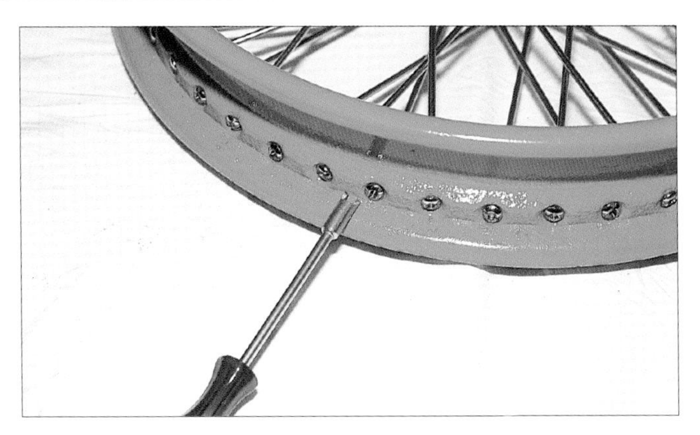

Gently does it: the nipples are slightly tensioned.

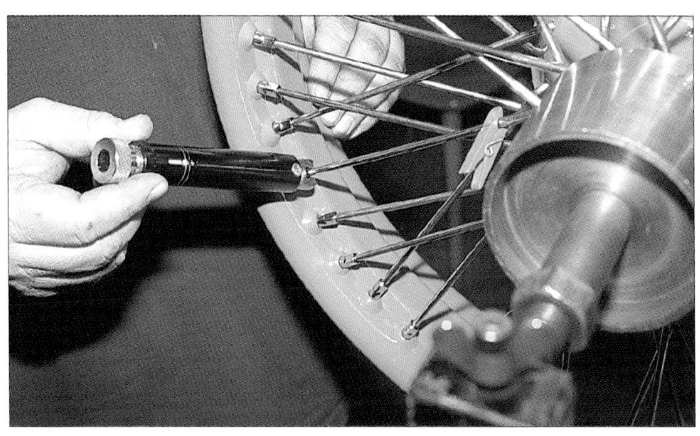

A torque wrench ensures the same torque for all the spokes.

Only the test rig will show how true the wheel is. Here, the lateral run-out is being measured.

The second gauge provides information on any potential radial run-out of the wheel.

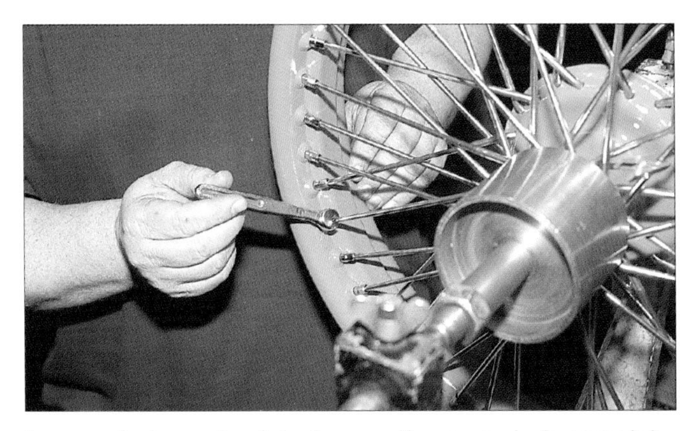

Even new wheels are not perfect: minor corrections are nearly always needed.

When the wheel is finished, the offset must be correct, or you will have to start all over again.

referred to as centring. While doing this, the wheel offset must be constantly monitored and adjusted as necessary. The ideal, of course, is a wheel which runs true and where all the spokes are equally tensioned. In practice, however, a degree of variation cannot be avoided, since tolerances apply even to new wheel rims and hubs.

Nonetheless, all the nipples must be tightened to within specified limits (this can be checked with a special torque wrench).

It all sounds pretty straightforward. "In principle it is, but practice makes perfect," says the specialist with a smile.

In conclusion, a few words about looking after wire wheels. It is most important to check all the spokes on a regular basis. Tapping them with a spanner or similar will immediately reveal a slack spoke on account of its duller sound. It should immediately be re-tensioned. The car should never be left for a long period with dirty wheels, so that the dirt is never given the chance to foster corrosion – a risk which can, of course, be ignored with stainless-steel spokes.

The splines on the hub will particularly thank you if you give them a light coating of copper paste. Come again? That all sounds very laborious. Well, perhaps it is, but take a look at the photographs. It's worth all the effort! And, in case you were wondering about the rather loud apple-green wheels illustrated, the MG J2 really was equipped with them when it was new ...

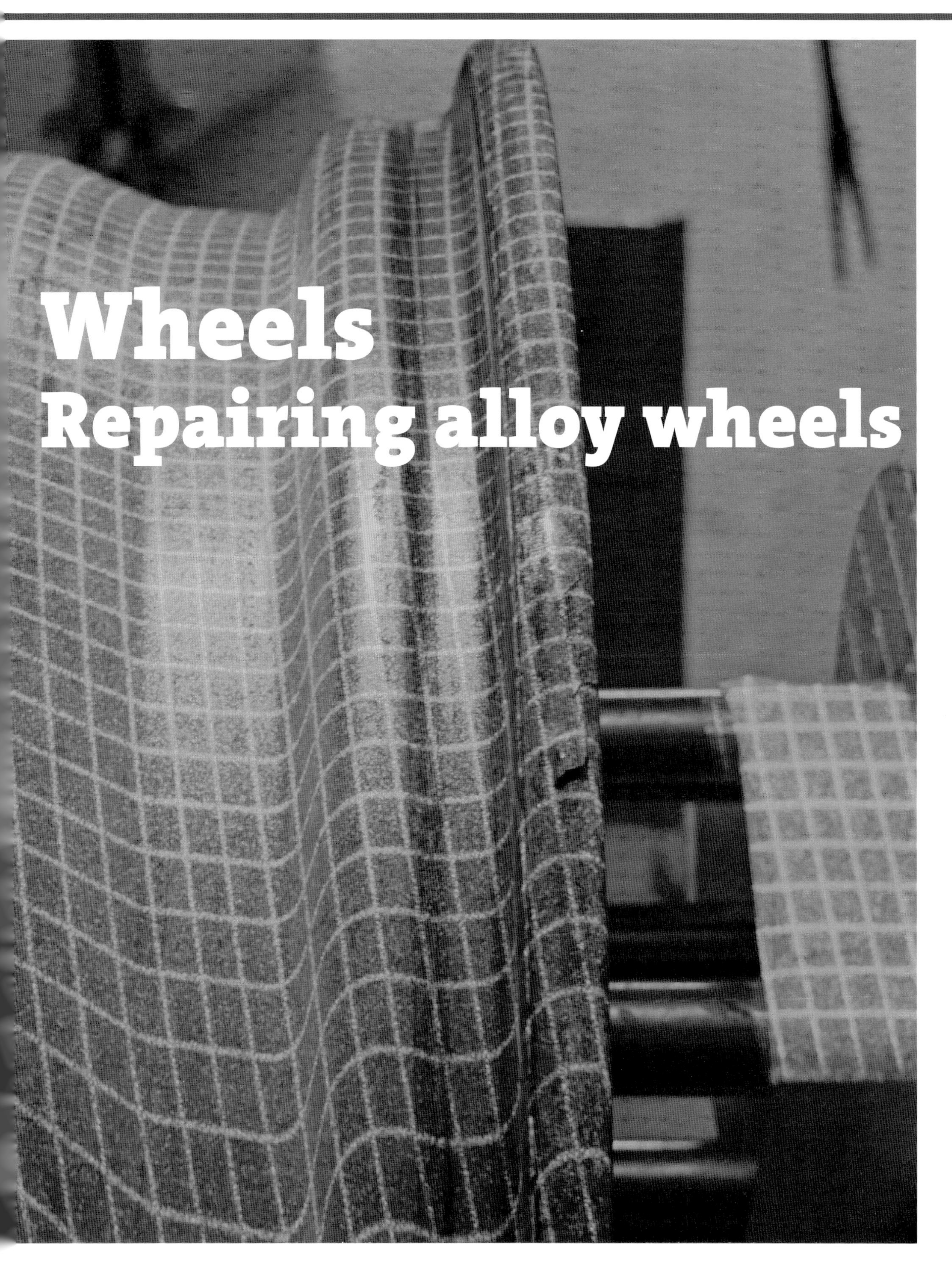

Wheels

Repairing alloy wheels

What should you do when the wheels on your car start to wobble? Repairing light-alloy wheels is a tricky subject. And not just technically, but legally, too: vehicle licensing regulations allow little leeway. We will show you what can be done, and what the vehicle examiners will accept ...

If potholes, kerbstones or mistakes when driving have ruined your fine alloy wheels, salvation is at hand, while staying legal – at least in the majority of cases.

One thing is clear: the word 'rim' strictly refers only to the outer section of the wheel, to which the tyre is fitted. This term dates back to the time when wheels were still made from several parts. For a long time, it was used even for one-piece wheel carriers with a solidly welded central section. We will stick to the generally accepted use of the term, even though multi-piece light-alloy wheels continue to be produced.

Light-alloy wheels reduce the unsprung weight of the car, which is beneficial not only in motorsport, but in everyday driving, when the lightweight wheels will make big, heavy saloons more comfortable to drive. And let's not beat around the bush, they certainly look better than steel wheels ...

The development of metal alloys and the production of aluminium and magnesium-alloy wheels has been refined over the years. Three-quarters of all the alloy wheels produced before 1980 or so would not pass the quality checks carried out by pretty much any manufacturer today: blowholes and other casting defects, or wheels which ran slightly out-of-true, were commonplace when alloys were new.

Light-alloy wheels first became fashionable as accessories in the 1960s. Soon, car manufacturers also offered them. The E3 (left) was available to order from BMW with BBS wheels. The Ronal wheels on the Glas 04 (right) were sold on the aftermarket.

To start with, the wheels are examined visually and to check that they run true.

On the inside of the rim, the specialist determines and marks the highest point of the imbalance.

The rim flange is also checked. Any small chips here can lead over time to the development of cracks.

An impact from a kerbstone, with slight distortion around it.

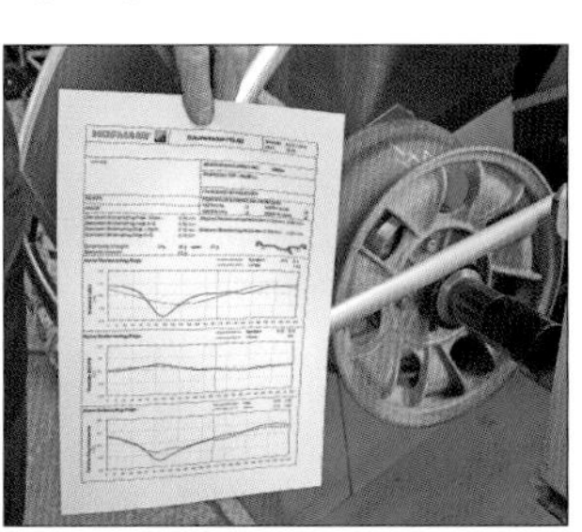

The final doubts are dispelled: the laser measurement reveals any impacts to the top or side of the rim.

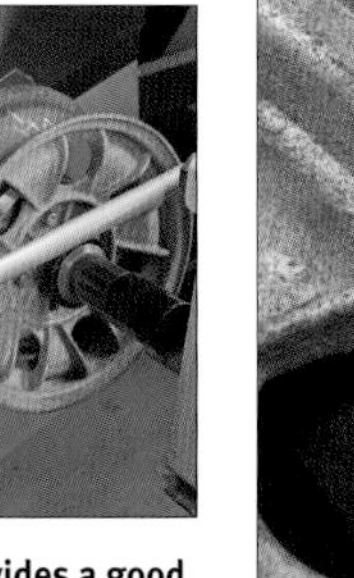

The inspection sheet provides a good deal of information: here it shows the presence of impacts to the top and side of the rim. The bead seat, on the other hand, is unremarkable.

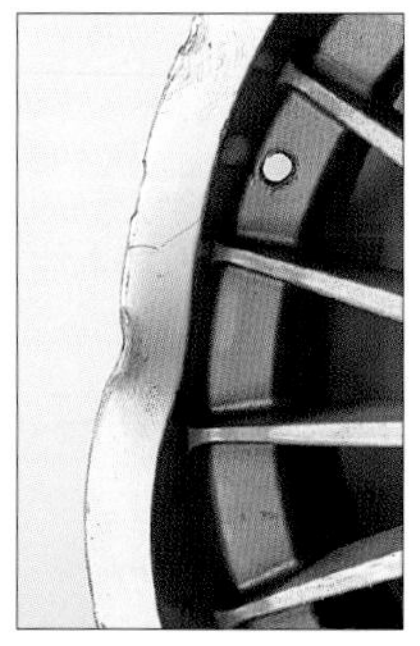

These wheels have had it: the fracture (on the left) has occurred as the result of attempts to repair the wheel. In the photograph in the middle, the distortion has gone too far down into the rim well; on the right, the damaged area is too large.

The 'roll-back' machine gradually restores the shape of the precisely tempered metal.

Right: The laser grid allows Stefan Mertens to check accurately that the wheel is running true. If it is, the gridlines projected onto it will be completely steady. If there is any imbalance, on the other hand, the gridlines will pulsate over the damaged area.

Perfectly true? One last test makes sure. The cosmetic restoration only starts once the wheel has been straightened.

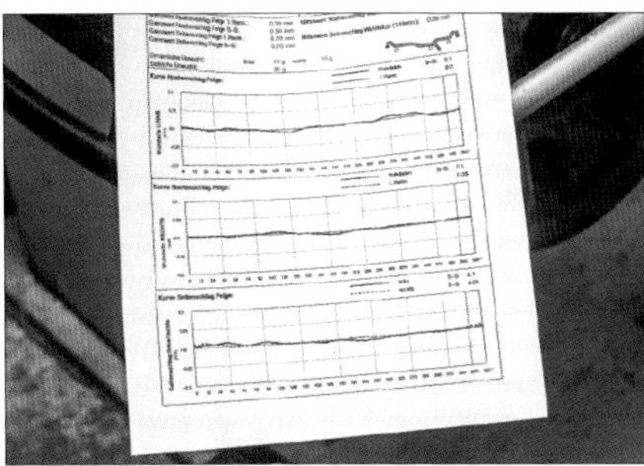

Close to new: the measurement diagrams show scarcely any deviation (see also page 135).

A deep impact from a kerbstone remains. This will be built up using laser welding.

The light-alloy wheels from the baby-boomer generation, however, have long since come of age on cars now officially recognised as historic vehicles. In the 1980s, wheels such as those from BBS RS, ATS' star-shaped wheels, or the 15-spoke alloy wheels made by Fuchs for Mercedes were all considered to be in good taste. Most of them went out of production many years ago.

So, what can you do in the event of damage, if the wheels cannot be found secondhand either?

The (reasonably) good news is that many types of damage (due to corrosion, potholes or kerbstones) can be repaired. Not all repaired wheels, however, are approved for use on the road. We put this to the test and visited Mertens Aluklinik (Alloy Wheel Clinic) in Mönchengladbach, with a set of shabby Lancia wheels made by FPS. We'll leave the restoration of rare wheels like these to the professionals.

Stefan Mertens has run his business since 1998, and offers a service to refurbish and repair all kinds of alloy wheel. What he can do, however, is tightly controlled: "Repairing wheels is permitted. Driving on repaired wheels on the public highway, however, is not allowed. Strictly speaking, kerbing damage to the edge of the wheel or corrosion damage can be sanded down or filled. Old coats of lacquer or powder coatings must be rubbed down by hand and carefully disposed of. Shot-blasting, welding, bending or hot alkaline cleaning – anything that could affect the structure of the wheel – has until now been

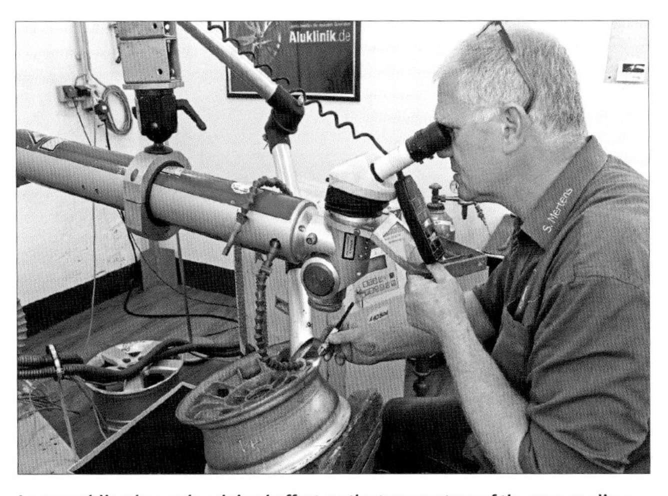

Laser welding has only minimal effect on the temperature of the surrounding metal. The distortion caused by heat absorption is only 0.7mm (0.03 in). Previously, TIG welding was used.

The fine strand of aluminium fuses onto the base metal in milliseconds.

A file is all that is needed to finish it off. The laser-welding process is absolutely exact.

The canon of beauty in the 1980s: spoilers and wide alloy wheels. The cars that have survived from this era are now becoming classics. And the rarest replacement parts today must surely be their period alloy wheels. Many of them were only produced for a short time, and with no replacements available, repairs can be the only way. Reproductions without a type approval number are not a legal alternative.

forbidden for use on public roads. Alloy wheels which have been welded or shot-blasted are only approved for use on racetracks or private roads." The reason for this is easy to explain: in Germany, at least, the statutory strength test carried out by order of the Federal Motor Transport Authority entails distorting and destroying the wheel.

Repairs are therefore practically out of the question, since testing that they had been properly carried out would require ruining the wheel once again ... The Alloy Wheel Clinic is, however, working to obtain approval for the use of repaired wheels on public roads – more on that at the end of this chapter.

"There's a lot we can salvage," said the specialist Stefan Mertens. Our FPS wheels were a good case study for different types of damage, and, unfortunately, for the limits professionals have to set on the work they do. Two of the wheels were eliminated at the very first stage, during the visual inspection, as the cracks in them made them unrepairable. "The cracks go down into the central part of the wheel, ie its load-bearing structure. What is worse, though, is that someone has already attempted to repair the wheel, and this has created the cracks. The rim flanges show signs of hammer blows, and next to them the metal has been fractured. We cannot accept any responsibility for wheels which have been messed about with like this. We have produced an in-house list of the acceptable limits for the damage we can repair. We exclude radial run-out values of more than 2.5mm (0.1in) or kerbstone damage exceeding 2.5mm (0.1in) in depth, or 20% of the total thickness of the rim flange."

All the same, the other Lancia alloys could be saved. Apart from kerbing damage, they only showed slight knocks. There were no cracks or deep pitting. The wheels were put on a balancing machine, and the imbalance recorded was within acceptable limits, though there was scope to do more. Using a gauge, Stefan Mertens tested the uniformity of the rim flange and the surface of the rim where the tyre was fitted. This enabled him to determine and mark any points which had suffered an impact.

Taking measurements by laser is also possible: the wheel is scanned in three places as it turns, with any imbalance displayed in a diagram. In the case of fine cracks or unusual signs of damage, the Alloy Wheel Clinic can also X-ray the wheels. The load rating of magnesium-alloy rims can be checked by means of a lateral pressure test, which the firm developed itself.

We now come to that stage which we can only illustrate with a single symbolic photograph: the 'warm roll-back process' – which is designed to protect the materials – is a trade secret of the Mönchengladbach firm. As the name indicates, the process 'undoes' the original distortion. No material is removed, as that would adversely affect the load-carrying ability of the wheel. The force applied by the machine and the temperature to which the wheels are heated are adapted to the characteristics of the alloy.

Off with the old lacquer. This is carefully carried out in the alkaline bath at 65°C (150°F). Steer clear of so-called 'high-temperature paint stripping processes'.

Alternatives to shot-blasting: the lacquer is carefully removed using very fine abrasives.

In our case, only a wafer-thin layer remained. Now, the wheel can be lacquered.

Powder-coating is an alternative to lacquering. For this, the wheels must be thoroughly dried after they leave the alkaline bath. When they were new, many alloy wheels were first sprayed with silver lacquer, and then received a clear powder coating. Curiously, both techniques are allowed, but the preparatory work is for the most part not approved for wheels used on the public highway. The Alloy Wheel Clinic is working to have the process made legal.

Nearly all minor chips or impacts to the top or side of the rim can be dealt with. Our specimen alloy wheels were restored to their correct shape in about 40 minutes, and checked again to ensure that they ran true.

Next came the kerbing damage. These sections were built up by welding on fine strands of the same metal by laser. According to our expert: "This process massively reduces heat absorption by the wheel. In the past, we repaired damage like this using TIG welding. Admittedly, we never received any complaints, but the heat absorption distorted the base metal by up to 20mm (0.8in). Using laser welding, this has gone right down, to 0.7mm (0.03in)."

Once the technical properties of the wheel have been restored, attention turns to their appearance. To remove old lacquer, the Alloy Wheel Clinic uses an alkaline bath heated to just 65°C (150°F). The wheels can be soaked in this without any risk of damage to the metal. "We recommend against using the high-temperature paint stripping processes at 400°C (750°F) which you sometimes see advertised. They may be quick and very thorough, but they affect the properties of the base metal, and, consequently, the load the wheel can bear," notes Stefan Mertens, urging owners to be patient.

Only wheels which are polished or have a very thin lacquer coat can be shot-blasted straightaway. For these, the Alloy Wheel Clinic uses very small stainless-steel balls at low pressure, in order once again not to affect the base metal. After this treatment, our two remaining Lancia wheels looked superb once more, and were ready to be lacquered or plastic-coated. The only fly in the ointment was that we unfortunately needed to find two more wheels to make up a complete set. Professionals set their limits where safety is concerned. The price? Reckon on about ●x70 to ●x140 per wheel.

For and against alloy wheel repairs – trying it for ourselves

The restoration described on the preceding pages was something of an emergency: the Lancia wheels are no longer available, and finding replacements can be very difficult. However, we also wanted to find out what a specialist can do in the case of wheels for a very popular modern classic which are still available. We took a set of Mercedes' well-known 15-spoke alloy wheels (with the classy, polished finish), which had been used over many winters on a W124 estate and understandably looked pretty 'tired.' We used these to find out what wheel repairs are really allowed these days. Or not, as the case may be.

In order to understand what wheel repairs are possible, consult your local wheel repair centre, and/or, vehicle testing authority (such as the DVLA in the UK). In Germany, for example, any 'abrasive treatment which entails removing any metal from light-alloy wheels is prohibited.' Processes, therefore, such as sandblasting or reshaping wheels using heat are strictly forbidden. Alloy wheels are granted type approval only after extensive testing, for structural integrity and load-bearing capacity. If approval is no longer valid the wheel can not be fitted to a car.

in line with standards set by the authorities, tested to destruction, that is, so there seems little point in attempting repairs of this kind.

Steel wheels, on the other hand, are generally more resistant, and can be stripped back and re-laquered if necessary, without compromising strength.

You should always take into account the forces to which a wheel is subjected when driving. When cornering, for example, when the weight of the car is transferred to the outside wheels, loads of 600-700kg (1320-1540lb) – and even more in the case of heavy cars – are at work, and the wheel must support these. It doesn't bear thinking what might happen if a wheel were to crack in these conditions. If it were later to emerge that a repaired wheel had significantly contributed to the causes of an accident, the law is unequivocal: the responsibility would be entirely yours! This fact is not lost on Stefan Mertens, head of the Alloy Wheel Clinic in Mönchengladbach, who had meanwhile assessed our Mercedes wheels and, to our surprise, came to the conclusion that – apart from their obvious visual faults – they had no serious damage. After a complete "cosmetic restoration," as he

Our Mercedes wheel had some nasty kerbing damage, apparently going back several years. The wheel should no longer be driven in this condition.

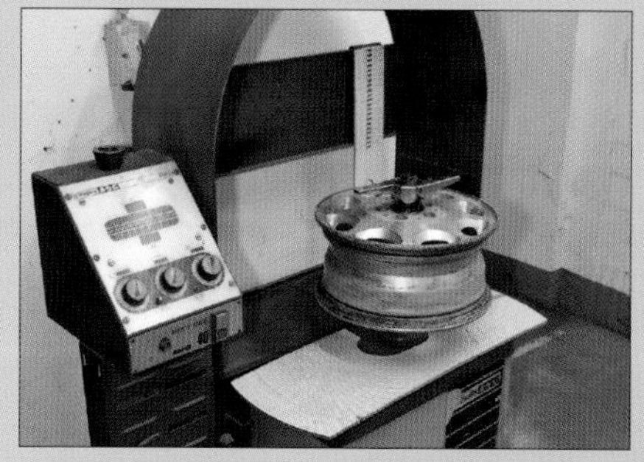

As part of the initial inspection, the wheel is subjected to a preliminary analysis.

This indicates that the distortion on the edge of the rim can be repaired.

The reading proves that the Mercedes wheel has taken a knock, but the damage is within the limits that can be repaired.

The manager of the Alloy Wheel Clinic, Stefan Mertens, uses a felt-tip pen to mark the damaged areas.

T1 denotes the main point of impact. Areas with 'collateral damage' are marked sequentially.

Traditionally, repairs to kerbing damage and cracks, for example, would have involved welding, but heat treating an alloy wheel constitutes an unauthorised process, as it's unlikely such a wheel would satisfy the TUV strength and load-bearing requirements. Furthermore, again in Germany, the TUV points out that any repaired wheel, in order to be legally certified, would have to be tested

termed it, he saw every likelihood that these sought-after modern classic wheels would look virtually as new. They could then be refitted without difficulty and would even be guaranteed.

The work on our Mercedes wheels began with a comprehensive preliminary analysis, together with a measurement of their geometry, which would indicate if

The test bench looks like a regular wheel-balancing machine, but can do much more ...

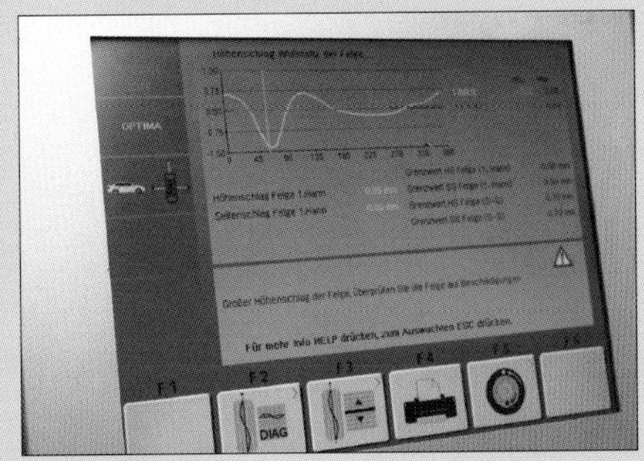
After just a few revolutions, it reveals that our Mercedes wheel has received a major impact to the top of the rim.

Before continuing the treatment, the wheel is carefully cleaned ...

... and washed down.

Next, it is placed in an alkaline bath. This is heated to a moderate 65°C (150°F), so that the metal is not affected.

The alkaline bath leaves the wheel in this rather insalubrious colour, but this disappears after it has been cleaned with a high-pressure jet.

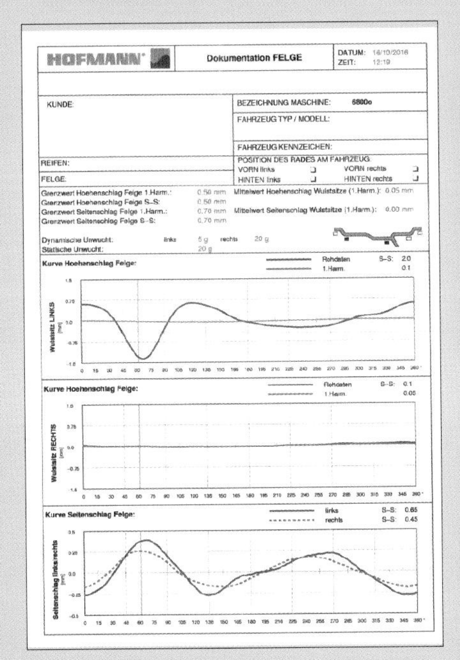
HOFMANN® | Dokumentation FELGE | DATUM: 14/10/2016 ZEIT: 12:19

KUNDE:
BEZEICHNUNG MASCHINE: 6800o
FAHRZEUG TYP / MODELL:
FAHRZEUG KENNZEICHEN:
REIFEN:
FELGE:
POSITION DES RADES AM FAHRZEUG:
VORN links ☐ VORN rechts ☐
HINTEN links ☐ HINTEN rechts ☐

Grenzwert Hoehenschlag Felge 1.Harm.: 0.50 mm
Grenzwert Hoehenschlag Felge S–S: 0.50 mm
Grenzwert Seitenschlag Felge 1.Harm.: 0.70 mm
Grenzwert Seitenschlag Felge S–S: 0.70 mm
Mittelwert Hoehenschlag Wulstsitze (1.Harm.): 0.05 mm
Mittelwert Seitenschlag Wulstsitze (1.Harm.): 0.00 mm

Dynamische Unwucht: links 5 g rechts 20 g
Statische Unwucht: 20 g

Kurve Hoehenschlag Felge: Rohdaten S–S: 2.0; 1.Harm. 0.1
Wulstsitz LINKS [mm]

Kurve Hoehenschlag Felge: Rohdaten S–S: 0.1; 1.Harm. 0.05
Wulstsitz RECHTS [mm]

Kurve Seitenschlag Felge: links S–S: 0.65; rechts S–S: 0.45
Seitenschlag links/rechts [mm]

The test report again shows the detailed condition of the wheel.

Before taking the wheel to the laser-welding room, the expert gently sands down the damaged section of the rim flange.

The centrepiece of the Alloy Wheel Clinic: in the laser-welding room, the damaged rim is repaired with minimal thermal stress for the metal.

For this, the expert welds the metal onto the marked-up position. It takes a while.

This repair technique is so kind to the base metal in part because the laser barely causes the edge of the rim to heat up at all. The spot which has been welded cools down so quickly that you can touch it immediately afterwards.

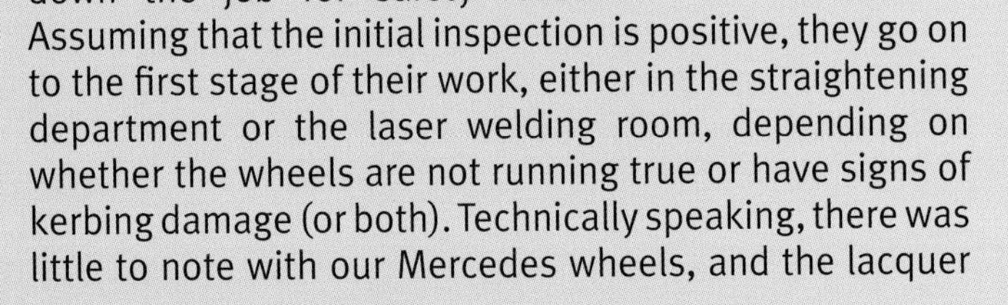
the wheels had previously sustained any damage which had not been professionally repaired. In this case, the Alloy Wheel Clinic would turn down the job for safety reasons. Assuming that the initial inspection is positive, they go on to the first stage of their work, either in the straightening department or the laser welding room, depending on whether the wheels are not running true or have signs of kerbing damage (or both). Technically speaking, there was little to note with our Mercedes wheels, and the lacquer was removed using chemicals at a temperature of 65°C (150°F); the wheels were then shot-blasted using small stainless-steel balls. After this so-called 'pre-conversion,' a layer of priming powder was applied, followed by a layer of silver powder and a final

The control panel for the welding equipment.

Clearly visible: the section which has been repaired leaves a small bead ...

... which disappears after the sanding machine is used once more.

Now it is time for the cosmetic restoration of our shabby-looking Mercedes wheels, which had seen service through many winters. First, they were shot-blasted ...

... and then sprayed with a clear powder coating, ...

... which was then baked on in the oven.

After this came a layer of silver lacquer, which set dry ...

... and was again baked on.

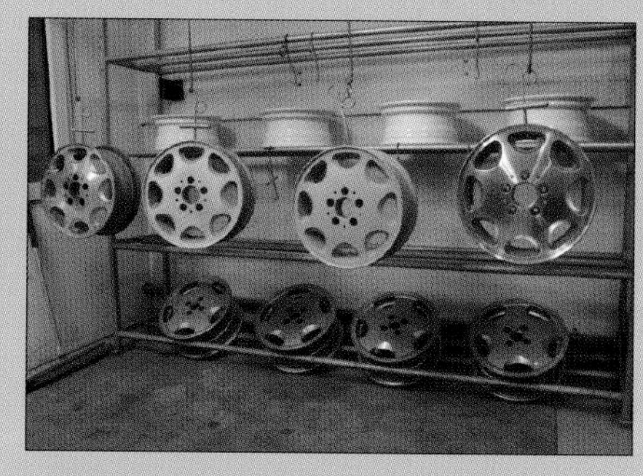

A perfect example of the different steps in treating our 15-spoke alloys: on the left, the shabby condition they started in, then with the two powder and primer coatings, and on the right the polished wheel at the end.

The laser grid shows how true the repaired wheels run. Although the wheel is spinning, the gridlines are steady, and the wheel is true.

The front of the wheel is prepared for the final stage ...

... which calls for some work by hand. And so the wheel receives its elegant shine.

clear lacquer coat. Following this, the Mönchengladbach repair firm treated the entire raised front surface from the centre of the wheel to the edge using a polishing process which it has developed itself, in order to restore the classy appearance of these German wheels.

Finally, the wheels were given unique markings and subjected to a further visual inspection and geometry check.

Regarding the cost: if you were to start with a decently maintained set of used wheels of this type, you might expect to pick them up for around •x450 on an internet auction site. Alternatively, you could drive down to your nearest official Mercedes-Benz dealership and order a

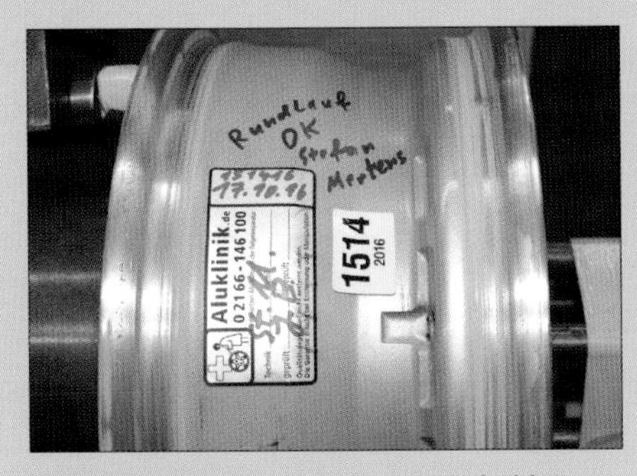

For the record: the repair is documented with indestructible stickers ...

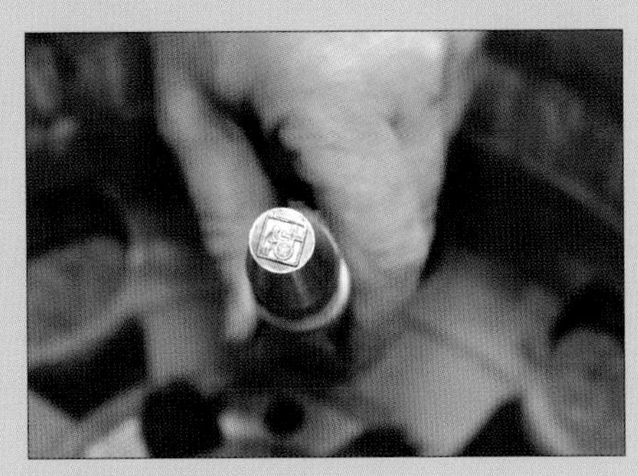

... and a die stamp.

Was there something there? It's true, there is no trace of the nasty knock, after the edges of the rim have been smoothed down again.

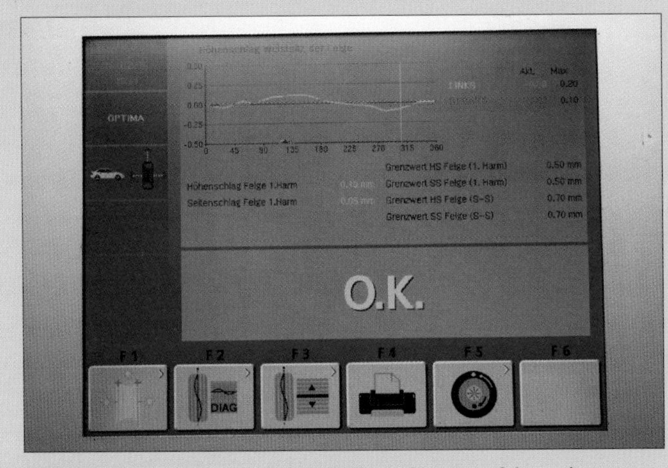

The laser measurement also confirms it: the performance of the ageing Mercedes wheels stands comparison with a brand-new rim.

Dr. Mertens in front of his clinic with one of our finished wheels.

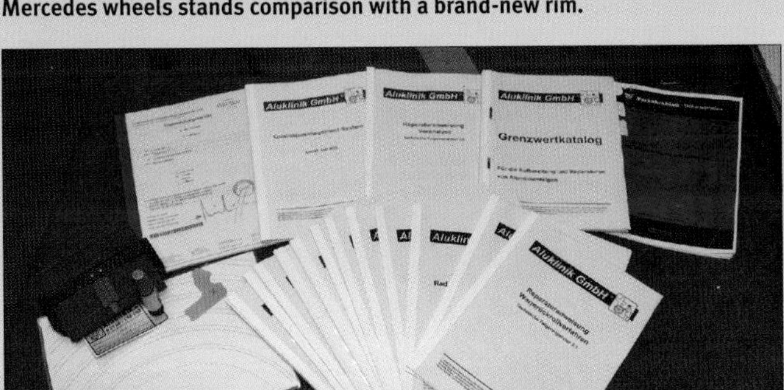

Pioneering work: the meticulous documentation accompanying each step in the work will (hopefully) lead to wheel repairs using the Alloy Wheel Clinic's system becoming legal.

new set. In the standard 7 x 15 size, the parts manager there, however, would charge you a not inconsiderable •x560. For a single wheel! Given this, a reasonably priced repair can prove worthwhile. If, as in our case, it is actually possible. The specialist reckons on an average of approximately •x135 per wheel to restore a set of alloy wheels. Since our Mercedes wheels needed a complete cosmetic restoration, the cost was about •x310 for each wheel, which is still very affordable when compared with the cost of the new items.

The final question remaining is whether that is actually allowed. According to the specialist Stefan Mertens: "We established many years ago that our repair methods have so little effect on the structure of the metal in the wheels we treat, that following a successful repair their strength remains within the tolerances set by the manufacturer. Using the laser-welding process we have developed ourselves, the deviations are even smaller. At the start of October 2016 we therefore submitted this process for approval to a working group of the specialist committee for automotive engineering. If the working group goes along with our approach, we are hopeful that the guidelines for wheel repairs in paragraph 36 of the German vehicle licensing regulations will be changed to authorise this type of repair." That would be great news for all classic car owners, who currently – not to put too fine a point on it – are driving on the edge of legality. But it is also often the case that there is no solution other than to repair or restore ageing wheels. What can you do if the wheel was very rare and production of it has long since stopped, or the manufacturer has simply gone out of business? Unlike our Mercedes wheels, there is no longer any hope of buying a replacement set. Even the secondhand market may be of no use. How long it will take before this change to the regulations is adopted, the specialist was, for the moment, unable to say. Customers who are interested will, however, find regular updates (in German) on his website at www.aluklinik.de.

Super-sicher ins Ziel
126
OLIO
Super-sicher ins Ziel
126

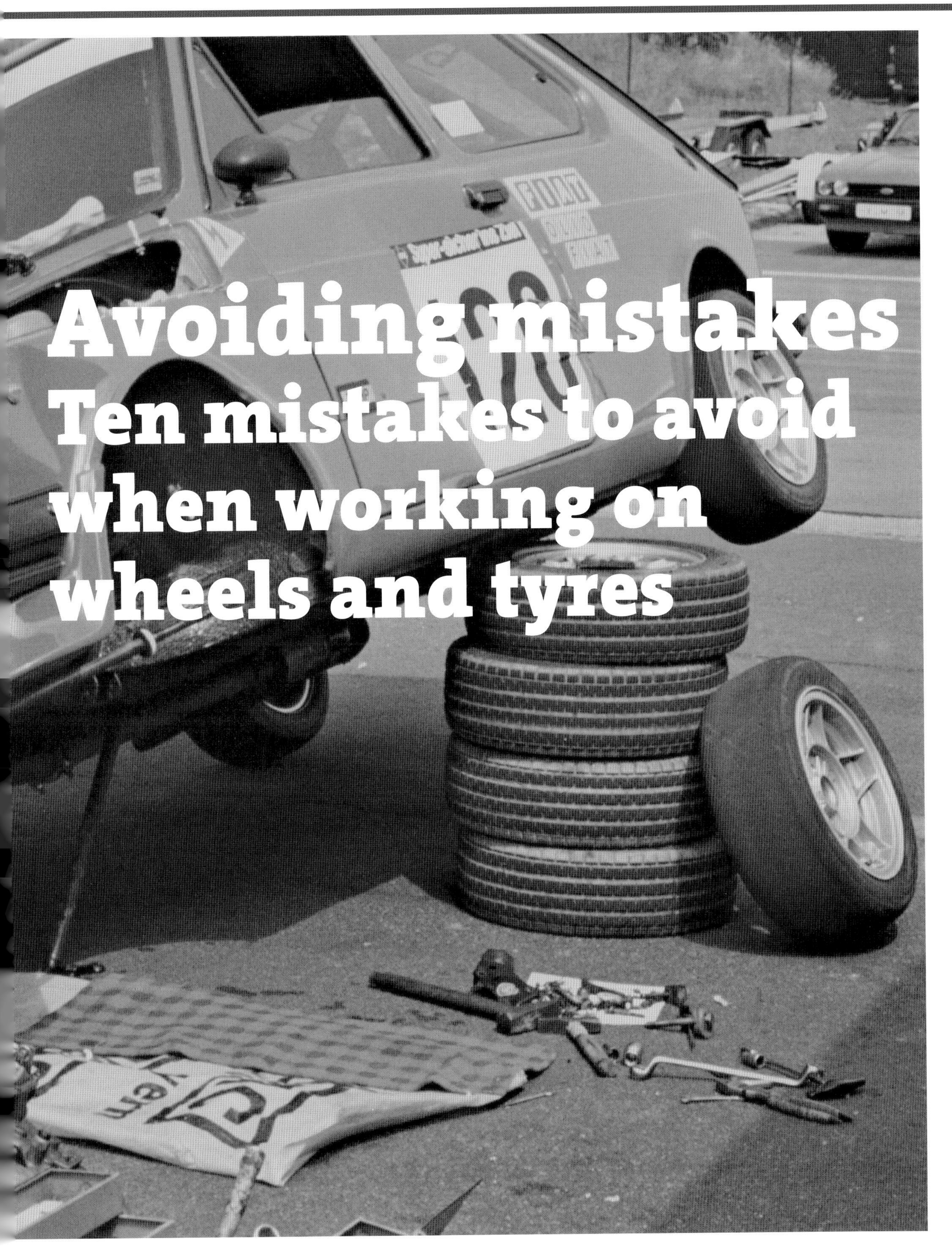

Avoiding mistakes
Ten mistakes to avoid when working on wheels and tyres

1 Greasing or lubricating wheel nuts when fitting them

Please don't! No kind of lubricant has any place on wheel nuts. Whether it's WD40®, Ballistol, or even the copper paste widely used by amateur mechanics. The reason is that lubricants of any kind reduce the frictional coefficient of the metal, which can lead to 'over-torqueing' the wheel nuts when fitting them, and, eventually to their shearing off. Clean the wheel nut threads with a wire brush instead, and put them on 'dry.' That is the only way to guarantee that they will be securely attached for the long term.

2 Driving with tyres which have different tread patterns

The widely-used term of 'mixing tyres' refers to the simultaneous use of crossply and radial tyres, which is prohibited! Since, however, even classic cars – with only a few exceptions – are now predominantly fitted with radial tyres, it is worth stating that it is quite possible to

Absolutely not! Lubricants have no place on wheel nuts.

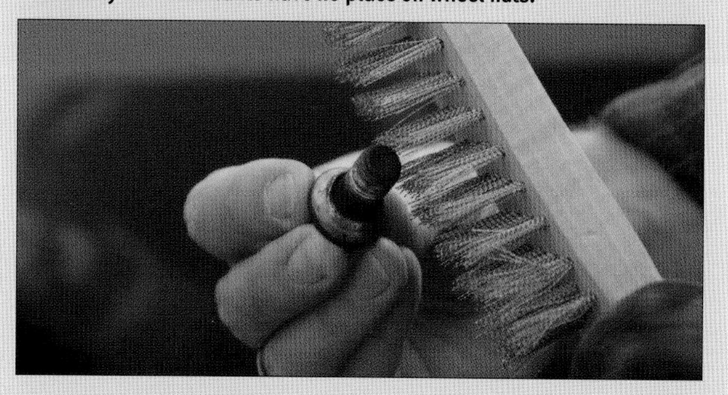

The right way to do it: cleaning the wheel nut threads with a wire brush before attaching them.

Although often recommended among amateur mechanics, copper paste should also never be used on wheel nuts.

Wheel nuts
Never apply oil or grease to wheel nuts.
Always replace damaged or corroded wheel nuts.
Never drive a car with a damaged thread on the wheel hub.
Clean the wheel and hub contact surfaces.
Only use OEM wheel nuts. Other types may work loose.

Even car manufacturers (here Mercedes) strongly advise against greasing wheel nuts.

Racing teams have the right tyres for all weather conditions, but never fit different tyres to the same car.

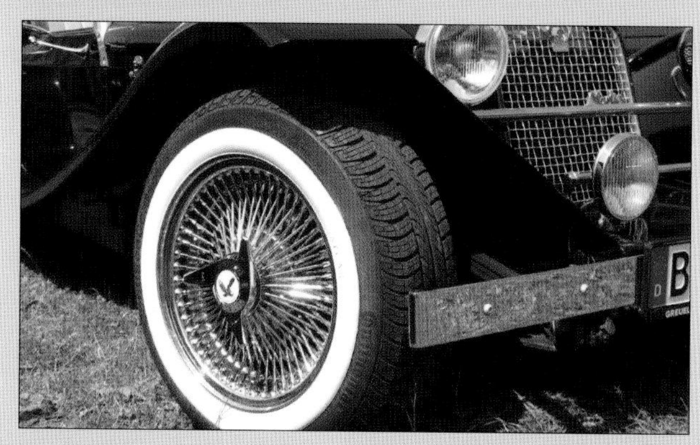

In actual fact, it is not forbidden to use tyres with different tread patterns, even fitted to the same axle. That applies equally to older classics ...

... and modern classics. But it is not to be recommended, as the car's handling can suffer considerably as a result.

use tyres with different tread patterns and from different manufacturers, even on the same axle. You can go so far as to mix summer and winter tyres, or tyres with different tread depths. None of this is expressly forbidden, provided nothing else is stipulated in the documents for that vehicle. You should always be aware, though, that you won't do the car's handling any favours, and, above all, that you shouldn't drive like this indefinitely. The law's main concern is that all the tyres should be of the same size and that they correspond to those specified in the documents for the car.

3 Driving with 'ancient' tyres, because "they still have plenty of tread"

For a long time, the story did the rounds that vehicle testing authorities would mark down tyres which were too old as a fault. And in the meantime, another story

The DOT number on the sidewall shows the week and year in which the tyre was manufactured (in this case, week 17 in 2010).

spread that this was in fact not the case. The age of the tyre is not in itself the basis for a fault! Inspection organisations are only interested in the condition of the tyres. If they are free from cracks, bulges or other visible damage, and are neither too heavily nor too unevenly worn, they may continue to be used. The following point should, however, be noted: as with all rubber products, so-called 'plasticisers' are added to tyres to keep the rubber flexible. After about seven to ten years, however, the rubber compounds have generally become so hard that although there may be no visible signs of wear, the tyres can no longer adapt to different driving conditions and will begin to slip. They are no longer really safe. For your own sake, therefore, you should check all the tyres regularly, and replace them on a preventative basis (at most) every ten years. Even if they still look alright.

4 Always replace damaged tyres

It depends on the damage! Vehicle licensing regulations govern the assessment of damage to, and repair of, car tyres; no method of repair is specifically excluded, so repairs are allowed. They should be left to professionals, though! In general, damage can be repaired in these circumstances:

This can be repaired. Leave the nail in place until you have reached the garage!

This tyre will have to be scrapped! A crack like this in the sidewall cannot be repaired.

Various car accessory shops and D-I-Y (Do-It-Yourself) stores sell repair kits to do the job yourself. However, ...

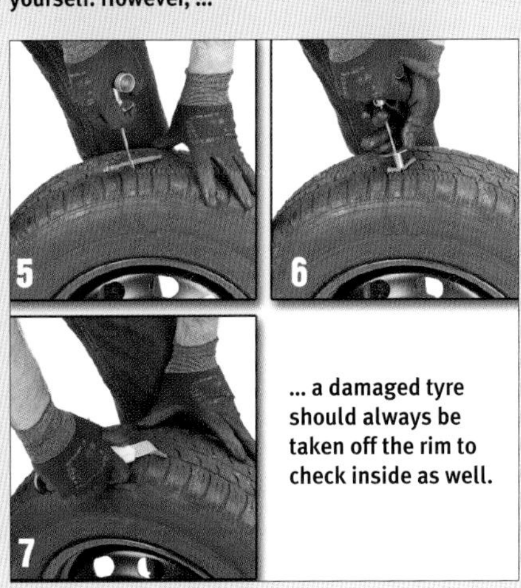

... a damaged tyre should always be taken off the rim to check inside as well.

- The damage to the tyre is discovered as quickly as possible
- The damage is to the tread and not to the edge of the tyre
- There are no large cracks (greater than 5-6mm/0.2in) or missing sections of tread
- The tyre is still recent

In general, in order to assess the damage, the tyre must be removed from the rim and the inside also examined. If the tyre in question has been driven for some distance with too low pressure, its structure may have been too greatly damaged for a repair to be possible. The same applies if an emergency repair has been carried out incorrectly. Please take note: if you drive over an object such as a nail, do not try to remove it, but leave it to 'seal' the hole in the tyre until you reach a garage. If the garage is farther away, check the air pressure regularly during your journey there and drive as slowly as possible. It is forbidden to repair a damaged tyre by putting a tube in a tubeless tyre!

5 Cleaning wheels with a high-pressure jet

We've all been there: after an enjoyable trip, possibly lasting several days, the wheels, especially at the front, are black with brake dust. It looks hideous and, if it is not removed promptly, it can even eat into the lacquer surface on the wheels. The solution is to clean the wheels using a high-pressure jet. But you should go about it with the utmost care. The hot water jet, which comes shooting out of the lance at high pressure, certainly removes the dirt, but can also badly damage the tyres. Always work at a good distance: the water jet can be lethal for the rubber!

Take care when cleaning your wheels at the car wash and make sure that the high-pressure jet is never aimed directly at the tyres.

6 Driving with over- or under-inflated tyres

Some people hang on stubbornly to the theory that increased tyre pressures lead to lower fuel consumption. To some extent, that is true, since the rolling resistance of the tyre is reduced. But this should not be exaggerated. If you over-inflate the tyres, only the centre of the tread will be in contact with the road surface, which will reduce grip

The tyre pressures should be checked every two weeks. Both under-inflating and over-inflating the tyres should be avoided.

and increase the stopping distance. It also increases tyre wear. A moderate increase in the tyre pressures is only recommended when the car will be driven with a heavy load (going on holiday, for example).

It is more common to find tyre pressures which are too low. This happens simply because at least two-thirds of all drivers fail to check their tyres every two weeks, as is recommended. In the long run, this is also unwise. If the tyre is under-inflated, only its shoulders – and not the tread – are in contact with the road surface: this causes the tyres to heat up and accelerates wear. It stands to reason that the reduced contact area also increases the stopping distance.

7 Driving with incorrectly tightened wheels

All car manufacturers provide the torque settings for the wheels they fit as standard. This information can usually be found in the owner's manual. You should stick to these settings. This is only logical: if the wheel nuts are not tightened sufficiently, the wheels could come loose when the car is driven, something we could all do without. If, on the other hand, the wheel nuts are done up too tight, in the worst case they could shear off, which could also ultimately lead to the wheel coming off. For this reason, we recommend that you always use a torque wrench with a clearly readable scale whenever you are fitting wheels.

8 Driving with wheels which are unbalanced or out-of-true

Basically, of course, you can drive with wheels which have not been balanced. But it isn't sensible to do so. Regardless of how carefully a wheel has been produced, once a tyre has been fitted to it, it will never run absolutely true. In order to compensate for this imbalance, the tyre fitter secures the wheel in a balancing machine, which spins it up and measures precisely in which places the wheel 'hops.' The fitter sticks or clips tiny weights to these places and repeats the procedure until the wheel shows no sign of any imbalance. Besides, it is not only